May 2/24

CORONA, CLIMATE, CHRONIC EMERGENCY

CORONA, CLIMATE, CHRONIC EMERGENCY

War Communism in the Twenty-First Century

Andreas Malm

VERSO

London • New York

First published by Verso 2020
© Andreas Malm 2020

The moral rights of the author have been asserted

1 3 5 7 9 10 8 6 4 2

Verso
UK: 6 Meard Street, London W1F 0EG
US: 20 Jay Street, Suite 1010, Brooklyn, NY 11201
versobooks.com

Verso is the imprint of New Left Books

ISBN-13: 978-1-83976-215-4
ISBN-13: 978-1-83976-217-8 (US EBK)
ISBN-13: 978-1-83976-216-1 (UK EBK)

British Library Cataloguing in Publication Data
A catalogue record for this book is available from the British Library

Library of Congress Cataloging-in-Publication Data
A catalog record for this book is available from the Library of Congress
Library of Congress Control Number: 2020941435

Typeset in Adobe Garamond by Hewer Text UK Ltd, Edinburgh
Printed and bound by CPI Group (UK) Ltd, Croydon CR0 4YY.

Contents

Corona and Climate

The third decade of the millennium began with the signing of another historic stimulus package for the dystopian imagination. Bushfires still roared through Australia, incinerating an area larger than Austria and Hungary combined, shooting flames seventy metres into the sky, immolating thirty-four humans and more than one billion animals, sending smoke all the way across the Pacific Ocean to Argentina and colouring the snow over the New Zealand mountains brown, when a virus jumped out of a food market in Wuhan, China. The market offered animals caught in the wild. There were wolf pups on sale, as well as bamboo rats, golden cicadas, hedgehogs, squirrels, foxes, civets, turtles, salamanders, crocodiles and snakes. Early studies pointed, however, to bats as the source of the virus. From that natural host, the virus would have jumped to some other species – pangolins were a prime suspect – travelled into the Wuhan market and made the leap into some of the human bodies circulating through the stores. Patients began to stream to hospitals. One of the first, an otherwise healthy forty-one-year-old man who worked at the market, spent a week with fever, tight chest, dry cough and various pains – during this week, coincidentally, temperatures in the afflicted Australian states sizzled above forty degrees – before being rushed into an intensive care unit.

Then the virus propagated through the world like a pulse through a grid. At the start of February 2020, some fifty people died every day, mostly from acute respiratory distress and failure, or not being able to breathe; by the first days of March, the daily global casualty toll stood at seventy; by the first of April, at 5,000, the exponential growth curve now nearly vertical. With at least one case of infection reported in 182 out of 202 countries, the pulse of death had crossed every ocean and streamed through streets from Belgium to Ecuador. And while this was happening, swarms of locusts larger and denser than any in living memory swept through east Africa and west Asia, covering the land, eating plants and fruits and leaving little if anything green behind. Farmers made bootless attempts to swat them away from fields. The clouds of locusts darkened the skies and, when falling dead, piled up in masses thick enough to stop trains in their tracks. One single living swarm in Kenya embodied an area three times that of New York City; a more normal contingent would have been one twenty-fourth that size and still contained up to 8 billion individuals, with an ability to devour the equivalent of what 4 million people eat in a day. Under normal conditions, any such swarms would be few and far between. The locust would stick to its solitary lifestyle in the deserts. But in 2018 and 2019, those deserts were showered with abnormal cyclones and torrential downpours, depositing such an excess of moisture that the eggs of the locusts multiplied into congregations so voracious as to threaten the food supplies of tens of millions of people, just as the virus descended.

No horseman of the apocalypse rides alone; plagues do not appear in the singular. It looks like there will be boils and thunder and pestilence and stinking rivers and dead fish and frogs in the kneading bowls. When these words are written, in the first days of April 2020, the total number of registered cases in the coronavirus pandemic is about to breach the one million threshold and the number of dead to pass 50,000 and no one knows how this will end. To paraphrase Lenin, it's as though decades have been crammed into weeks, the world spinning in a higher gear, leaving every forecast liable to embarrassment. But if we cash some of the cheques written to the imagination, we can envision a fevered planet inhabited by people with fevers: there will be global heating plus pandemics, slums sinking into the sea with people dying from pneumonia in, for example, Mumbai. The slum of Dharavi just reported its first case of the coronavirus. One million people live in close quarters in Dharavi, with minimal access to sanitary facilities, and every year storm surges invade the slum with higher waters. There will be refugee camps where pathogens eat their way through crowded bodies like a fuse through a wire. It will be too hot, and there will be too much contagion, to step outside. Fields will crack under the sun with no one to tend them – but on the other hand, the corona crisis came from the start with the promise of a return to normality, and this promise was unusually loud and credible, because the malady seemed far more external to the system than, say, the crash of an investment bank. The virus was the epitome of an exogenous shock. It would fizzle out, one month or the next. There might be a second wave but still that would be it. A vaccine could choke off the pandemic. Every measure taken to contain it was advertised as temporary, like police tape marking off a street, and so we can just as easily envision a planet lifted back to the *status quo ante*. The streets will fill up again. Shoppers will throw away their face masks with relief and throng the malls. There will be a pent-up urge for everyone to pick up where they left off when the virus struck, and it will be released with gusto: airliners back in the sky, their canopy of white contrails budding as if after winter. Private consumption might be more alluring than ever. Who would want to stand on a packed bus or train after this? Unutilised capacity in car and steel and coal plants will burst forth and stockpiled inputs fall in line with the supply chains. Out of sight, the oil drills back in operation, hammering away.

But these two opposing scenarios of the future are, on closer inspection, exactly one and the same.

Can there also be a way out?

Where an emergency exists

To hinder or at least slow down the spread of the virus, states across the globe took extraordinary – in every sense of the word – measures to confine their citizens to their homes. Lockdowns were implemented with various degrees of policing, some draconian. European prohibitions covered activities ranging from mingling with more than one other individual (Germany) to leaving the house without a permit (France) to leaving the house without a parent if you're under eighteen (Poland) to crossing into another municipality (Italy) to picnicking in a park, popping into a pub, dining in a restaurant and receiving foreign visitors (most countries). By early April, the generality of the human species fell under some version of shutdown. Never before had the business-as-usual of late capitalism been so utterly suspended.

All efforts were devoted to combating the pandemic, a distinction laid down between 'essential' and 'nonessential' functions in society. The palatial luxury store Harrods in London fell in the latter category; having stayed open throughout the bombings of the city in World War II, it closed on 20 March 2020. 'Starbucks is not essential', asserted one barista in Philadelphia who signed a workers' petition to shutter all stores in the US. A suffering pioneer, Italy ordered all 'non-essential' factories and businesses to close, exempting only outlets like supermarkets, pharmacies and post offices. The principle was unheard of: some forms of production and commerce meet basic human needs, while others have no legitimate claims to uninterrupted revenue streams and can be discontinued forthwith.

It followed that some things ought to be produced rather than others. One of the industries most blatantly non-essential to the moment was car manufacturing, and any factories risked serving as hothouses for the virus, and so, by mid-March, the auto giants of the world, from Volkswagen to Honda and Fiat Chrysler, switched off their assembly lines and sent workers home. Just-in-time supply chains had anyway been disrupted. But car manufacturing is also a wonder of technological prowess, with an unrivalled ability to mobilise robotics and engineering and shop-floor skills in the execution of novel tasks, assemble components in alternative combinations and churn out state-of-the-art commodities in no time, the conversions during World War II a reminder of that possibility. This time, the need was not for tanks and bombers, but for things like ventilators: machines for pumping air into lungs and sucking out secretions, so as to keep critically ill patients breathing. The US president Donald Trump was initially not fond of the idea – 'we're not a country based on nationalizing our business', he declared – and the US Chamber of Commerce resisted it, but eventually Trump found himself invoking the Defense Production Act, which allows a president to *command* private enterprises to provide essential goods in a crisis. GM and Ford began clearing out idle plants of redundant equipment, sourcing the requisite parts and figuring out how to mass produce ventilators at a pace to match the exploding pandemic. GM vowed to forego the pursuit of profit.

Having likewise sunk into the hole of non-essentiality, fashion brands like Prada, Armani, Yves Saint Laurent and H&M converted some of their manufacturing capacity to goods the health care sector cried out for: medical overalls, face masks, hazmat suits. No more cropped boleros or leopard-print suede boots. Distilleries from California to Denmark modified their vodka and whisky lines to instead deliver hand sanitizers. There was planned relocation of workforces: in Sweden, flight attendants from the grounded Scandinavian Airlines were re-educated as nurses and shepherded into the hospitals, reportedly expressing mass enthusiasm for the mission. No more duty-free perfume and jewellery in the aisles. Now it was all about saving lives.

In the emergency, the fences around private property blew away like a thatched hut in a hurricane: Spain nationalised all private health facilities in one stroke and instructed companies with potential capacity for producing medical equipment to align with state plans. Britain all but nationalised its railway system, while the Italian state took over flag carrier Alitalia. No other department of capital accumulation fell with as roaring a thud as aviation. By early April, half of the world's aeroplanes had gone into storage, London Heathrow shutting one of its runways and Boeing its factories, mass flying relegated to the era BC (Before Coronavirus). The most recklessly wasteful of all sports,

Formula One, went down. The Geneva Motor Show was cancelled. The giant annual CeraWeek gathering of oil and gas executives in Houston was called off: fossil capital entered a state of paralysis. As demand caved in, oil producers closed their rigs and wells, prices no longer enough to cover the costs of non-conventional production in particular. Fracking neared a standstill. ExxonMobil announced that it would slow down its developments in the Permian Basin of the southwestern United States, the El Dorado of shale oil and gas; across the board, two thirds of the investments anticipated for 2020 in new infrastructure for extracting oil and gas worldwide were shelved. 'Not only is this the largest economic shock of our lifetimes, but carbon-based industries like oil sit in the cross-hairs', Goldman Sachs commented. 'Accordingly, oil has been disproportionately hit.' Other analysts claimed that the oil sector faced the worst crisis in a century, which meant, effectively, the worst ever.

And so emissions plummeted. China, scene of the original outbreak and the world's largest plumes of CO_2, was the first to clear the skies. During February 2020, the combustion of coal dropped by more than a third, of refined oil products by slightly less; one company reported petrol sales falling by 60 per cent and diesel by 40. Domestic flights plunged by 70 per cent in two weeks. In the aggregate, China's CO_2 emissions fell by *one quarter in a single month*, a contraction swifter than anything ever witnessed but bound to be replicated as the pandemic and the countermeasures rolled around the globe, although the extent of the dip remained, like everything else, uncertain.

All of this was pervaded by a rhetoric of war. Heads of states cast themselves as commanders-in-chiefs of nations on a warpath – 'we are at war', proclaimed Emmanuel Macron of France; 'we're at war and we're fighting an invisible enemy', Donald Trump; 'we're at war and ventilators are our ammunition', Bill de Blasio, mayor of New York City, epicentre of the US outbreak. Parallels with World War II imposed themselves. 'Call it war mobilization', the *Los Angeles Times* commented on the ongoing conversion of production. Time had run out for normal politics, as even the most reluctant seemed forced to accept. 'Where an emergency exists and it's very important that we get to the bottom line and quickly, we will do what we have to do', admitted Trump. Ten days before he himself tested positive for the coronavirus and retreated into self-isolation and, eventually, hospital, Boris Johnson, prime minister of the UK, gave a press conference in which he solemnly vowed to 'act like any wartime government', with 'a profound sense of urgency'. Staring fate in the eye, he acknowledged that 'yes, this enemy can be deadly, but it is also beatable. And we know that if we follow the scientific advice that is now being given, we will beat it.' To some ears and eyes, this drama bore certain resemblances to a television script never aired.

We have an enemy out there

'I want you to panic,' repeated Greta Thunberg when she toured the halls of world politics in 2019. Leaders of many stripes – though not all – basked in the light of her rectitude and tried to capture her on selfies. But the one thing they did not do was panic. Nor did they take to heart the proposition that the climate crisis constituted an emergency on a par with war. For a number of years, that had been a staple of the agitation of climate scientists and activists, who liked to cite the Allied war effort as an actual case of society facing death, corralling its forces to survive, focusing on one aim to the exclusion of everything else and managing, under extreme time pressure, to defeat the enemy. The most cited paper on how the US economy could replace fossil fuels with 100 per cent renewable energy pointed to the factories of GM and Ford rolling out hundreds of thousands of aircraft during World War II. Then why not wind turbines and solar panels? In 2011, the year when that paper was published, NGOs spanning the gamut of the environmental movement urged the heads of the US and China to get on a war footing – after all, the WHO, the highest authority on matters of human health, estimated that global warming was already killing more than 150,000 people per year.

Few ministers and other top 'policy-makers' in the global North could have failed to hear of the parallel. In the most detailed comparison to date, *Strategies for Rapid Climate Mitigation: Wartime Mobilisation as a Model for*

Action, Laurence Delina, a sustainability scholar now based in Hong Kong, parsed the lessons for how states can marshal their resources – money, labour, technology – and phase out fossil fuels at the speed required. One reader of that book, Bill McKibben, the most famous climate activist until Greta Thunberg went on strike, brought his rhetorical flair to the analogy in a 2016 essay, 'A World at War', in which he described the latest season of Arctic meltdown as a devastating enemy offensive and the firestorms and droughts then in the news as overwhelming assaults, only to annul the metaphor: 'It's not that global warming is *like* a world war. It *is* a world war. Its first victims, ironically, are those who have done the least to cause the crisis. But it's a world war aimed at us all', and then he made the case for retooling the apparatus of production as in the previous world war.

McKibben was a key supporter of the Bernie Sanders presidential campaign in 2016. Sanders recommended that the US 'approach this as if we were at war' – 'we have an enemy out there' – but even though he lost the nomination, the Democratic Party officially adopted his demand for warlike mobilisation before the election. Hillary Clinton pledged to furnish the White House with a 'situation room just for climate change' modelled on the map room where Franklin D. Roosevelt managed the war campaigns. One offshoot of the US climate movement developed a 'victory plan' with the iconic silhouette of American soldiers hoisting not a flag but a wind turbine; demonstrations could be seen carrying front banners with

the text 'World War I, World War II, World War CO₂'
and the stern face of Uncle Sam. The generation of activists that came to the fore in 2019 stayed with the trope
– Alexandria Ocasio-Cortez among them – and again
pushed it up the ladder to establishment figures like Joseph
Stiglitz and Ed Miliband, both of whom called for a
wartime response in that year. And indeed, climate emergency and the need for panic were the leitmotifs of 2019
BC, confirmed when *Time* portrayed Greta Thunberg as
'person of the year', standing on a cliff as a wave crashed
against it, on the cover of the 23 December issue. On that
day, the forty-one-year-old worker from the Wuhan
market was still at home shivering and spitting.

How corona and climate differ: first cut

In this chamber of echoes and analogues, a question had
to arise: why did the states of the global North act on
corona but not on climate? More precisely, why did they
at best pay lip service to the ideal of doing something
about emissions, and then shied away from any measures
– not even putting their populations under house arrest
– to repulse the coronavirus-associated disease formally
designated Covid-19 by the WHO? The question was
much discussed on the online forums to which humanity
was condemned in March. A great number of explanations for the disparity were posted. The first to consider is
the claim, to which someone like Donald Trump would
presumably subscribe, that only one of the two problems

actually existed. That we can put aside. There is a related belief that Covid-19 represented a more serious danger to humanity in some absolute, objective, clinical sense. This is not tenable. Unabated global heating will burn through the foundations of human life (not to mention that of other species without number). Covid-19 could do no such thing; even if it were to make a quantum leap and reach Black Death proportions, killing off half of the population in Europe or some other continent, it would come to a halt at a safe distance from that terminus. In early April, scientists still debated whether the mortality rate for infected patients stood below 1 per cent or possibly as high as 10. Climate breakdown promised no such capacious exit clauses – it's not something from which a majority can recover – and so the sheer magnitude of the danger can hardly explain the unparalleled rush to war.

Nor can the state of the science. In fact, at the moment when governments sprang into action, the science of Covid-19 was riddled with uncertainties about virtually every aspect of the disease – whether the virus could travel through the air, whether it could be spread by people without symptoms, whether the recovered were immune, why deaths were so few in some places and so many in others, what strategies would work best to roll it back (contact tracing, mass testing, herd immunity, universal quarantine?). There was nothing like the consensus and coherence and decades of peer-reviewed literature and libraries filled with reports corroborating the fundamentals of climate science over and over again. Nations acted on a headline while ignoring a century of knowledge (which is not to say that the former was a bad move).

Someone suggested that 'with greenhouse gases you can't see them; you can't smell them', but no one ever spotted a coronavirus bumping along the streets or caught the stench of it. Some argued that the pandemic was invigoratingly simple. The 'bewilderingly complex' climate labyrinth pulled initiatives for action into an impasse. Everyone understood how the virus spread – through close contact between people – and the WHO provided 'immediately actionable paths' to mitigating it, with no equivalents in the domain of climate. But the mechanisms of anthropogenic climate change have never been more enigmatic than the droplets transmitting Covid-19 from one person coughing or sneezing onto another. Nor have the policy recommendations from the IPCC (Intergovernmental Panel on Climate Change) and other scientific agencies been very hard to comprehend: stop the emissions. Simple as that.

Some proposed that global heating remained a 'far-off probability', a 'distant, non-certain threat'; Covid-19, to the contrary, 'is not a problem for future generations, but for everyone living now'. As a statement of fact, the former was last valid perhaps 40 years ago. By March 2020, the WHO, to which governments now lent their ear, had been counting more than 150,000 annual deaths from climate change for *four decades running*, the curve of suffering and loss steadily on the rise. The World Meteorological Organization (WMO) estimated that 22 million people were displaced by extreme weather events in 2019, up from

17 million the year before. A catalogue of present-day climate misery from Uruguay to Korea, the WMO report for the last year BC enumerated lethal heatwaves, ruinous floods, calamitous cyclones, destruction of crops by crippling droughts and a spike in the cases of dengue fever, topped off by the wildfires in Australia and the locusts on the Horn of Africa. There is nothing probable or hypothetical about any of that. Equally faulty is the assertion that no specific weather event can be attributed to global heating – which was the state of the science sometime before the millennium – as well as the argument that while a pathogen makes for a perfect foe, the other candidate for a war has 'no single, clear "enemy." Who is to blame for climate change?' But the climate movement has for some time now followed the scientific clarifications and singled out the forces to blame. The enemy is fossil capital.

All of these explanations – the unreality of the climate crisis, its comparatively benign character, or uncertainty, intangibility, complexity, remoteness or lack of front lines – should be sorted under the rubric of ideology. They do not pertain to factual properties of the phenomenon, but to distorted perceptions of it. Their truth content consists in voicing facets of the ideology that *in itself* impedes action on climate: it's not the case that there is no enemy, but the belief in its absence contributes to passivity. The same applies to the view that 'the future is going to be bad regardless of what steps we take now to address climate change. This can beget feelings of helplessness. With coronavirus, it feels as though today's actions will have real and demonstrable consequences' – evidently the *effect* of the distribution of actions, not its cause. If governments had acted resolutely on climate, there would have been plenty of hope about it; if they had let the virus run amok, surely despair would have set in. The hopelessness resulting from the historical abdication of states in the face of danger can then feed on itself, validating a surrender to fate – but only if it is actively affirmed and reiterated. Decades of hard work from the enemy side have led up to these impressions.

But then there are also hypotheses that merit closer consideration, such as climate change being *gradual*, whereas Covid-19 was as sudden an explosion as the world has ever seen. This counted as among the more popular explanations in March 2020, the month of the great divergence. Here is a factual property of global heating: it does not appear out of the blue and then retreat to wherever it came from, as Covid-19 was expected to do. Gradualness, however, may not be the appropriate term for the quality. Climate breakdown could instead be seen as a landslide that rolls through the entire earth system, sweeping up material and gathering speed, and every time it hits people standing in the way, the impact is anything but gradual – if by 'gradual' we mean a slow, incremental accretion of factors that makes moment X on a timeline difficult to distinguish from moment Y. There is, on this definition, nothing gradual about a hurricane that knocks down all infrastructure on an island. An Australian inferno or a Kenyan locust attack does not feel just like, or nearly just like, another normal day. The blow can be shattering.

Some aspects of global warming, such as sea level rise, may come across as gradual in the given sense, but then again it often makes itself felt rather like a water bomb. In November 2019, Venice was hit by a deluge nearly two metres high, smashing through marble and brick, piazzas and basilicas in scenes of 'apocalyptic devastation', which the mayor of the city, like every other person not committed to denial, knew to pin on climate change. This appears to be the general form of the process: an avalanche of missiles, standing out from a singular occurrence like Covid-19 not by being gradual but by constituting *a secular trend that persists over decades and centuries* unless brought to a stop.

Gradualism – the expectation that global warming would obey the laws of linear causality, ticking evenly like a clock, adding one infinitesimal stressor to the next, in accordance with the hoary old dogma of *natura non facit saltum* – has long been a pernicious ideological filter, torn apart by scientists observing the moving object on the research frontiers. This is not the place for an extensive review of its demise. Suffice it to note that so fast are developments now that several new discoveries were publicised during a few days in March 2020: it turned out, for instance, that 600 billion tons of ice crashed into the ocean from Greenland the previous summer in only two months of record heat. That single dump translated to 2.2 millimetres of global sea level rise – so much more ammunition for the next surges. One analysis revealed the polar ice caps to be melting seven times faster than in the

1990s. On 10 March, *Nature Communications* presented one more nail in the coffin of gradualism, an empirical and modelling study of just how speedily regional ecosystems like the Amazon rainforest and the Caribbean coral reefs may come crashing down – over perceptible human timescales of years and decades – ending with a call to prepare for changes 'faster than previously envisaged through our traditional linear view of the world'. This is more like the opposite of gradual: a cascading series of abrupt destructions.

But still, there seems to be something in the respective temporal profiles of Covid-19 and climate change that accounts for the antithetical reactions. What exactly? When pondering this question, one should keep in mind that the *explanandum* is not the different *popular* reactions: it wasn't the French or British or Australian people that came together and decided on a lockdown – it happened way too fast for any democratic deliberation – nor did they take the decisions to postpone meaningful emissions cuts into the meaningless future, a policy about as receptive to mass demands as that of austerity. We are looking for an explanation as to why *states in advanced capitalist countries* got so relatively fired up about the virus. Here one observer pointed out that the victims of Covid-19 – at an early stage of the pandemic – appeared to belong to a core constituency of the ascendant right: old white people. 'Unlike the climate crisis, the virus predominantly threatens the elderly – the right's core support group – rather than millennials', and no government would survive the next election if they abandoned them to the devil. This would seem to be part of the story.

Once again, however, it is more complicated, for old white people have not gone unscathed from global warming. In 2003, Europe boiled in the hottest summer thus far; the peak of the heatwave killed an estimated 30,000 people, half of them in France. Over the course of twenty August days, one segment of the French population succumbed in droves – namely people over sixty-five with pre-existing medical conditions, victims of a mortality event robustly attributed to the underlying trend. Two heatwaves in the summer of 2019 ended some 1,500 French lives, again primarily among the elderly. Smoke from the bushfires entered lungs across eastern Australia; during nineteen weeks, it finished off more than 400 people in the four worst affected states, many of them elders with cardiorespiratory conditions, who were no longer able to breathe. Harmful particulate matter has other sources than wildfire, of course. When the blanket of pollutants covering cities like Wuhan and Shanghai was rolled up in February 2020, the air became so much easier to breathe that in the calculation of one Stanford researcher, the lives of 4,000 children under five and 73,000 adults over seventy were spared. That would have amounted to some twenty times more lives saved by the clean-up than lost to Covid-19 itself. An unintended side effect was a far greater rescue operation. At pains to pre-empt the interpretation that pandemics are panaceas, this researcher merely wanted to highlight 'the often-hidden health consequences of the status quo' – 'our normal way of doing things might need disrupting'. Bespeaking a degree of absurdity, some predicted the same balance – more lives saved by preventing air pollution than by preventing infection – in European countries, including France, where one researcher dryly noted that 'these are quite fascinating times'.

But then the very mitigation of the pandemic kept mortality down. No one knew, of course, just how deadly it might become. One estimate in late March suggested that absent interventions, Covid-19 would have killed 40 million people during 2020 – a truly terrifying number (more than five times higher than the annual toll from air pollution). Governments faced the prospect of hundreds of thousands or even millions languishing and expiring in their beds. By early April, it had become obvious that potentials for such mass death were largest in the global South, in places like Dharavi, to which Covid-19 had yet to arrive in full force. By then, several thousand had already died in Italy, Spain, Britain, the US. And here might be a key to the divergence: what we might call *the timeline of victimhood.* Just as Bill McKibben remarked, the war waged by fossil capital, which kills through the medium of the atmosphere, reaps its first victims among 'those who have done the least to cause the crisis' – poor people in the global South, that is. It might ultimately be 'a world war aimed at us all', but the rich will be the last to pass away. For Covid-19, the timeline was exactly the reverse: first to pass away on some scale were people in the richer North. We can pinpoint with accuracy the moment when the virus mutated into a genuinely global crisis, and it was not when it swept through Iran: it happened when hundreds began to die in Italy, more precisely the affluent northern province of Lombardy. That was when panic gripped Western governments. To make matters worse, a gallery of celebrities and politicians fell sick – Tom Hanks and his wife, Plácido Domingo, Kristofer Hivju, Patrick Devedjian, Rand Paul, Harvey Weinstein, Prince Charles, Prince Albert II of Monaco, the secretary general of the Spanish far-right party Vox and, of course, the prime minister of the UK – while a special place in the feverish hell was reserved for luxury cruise ships. None of these individuals or entities had faced much risk from the climate crisis. Nor had the IPCC declared Europe its 'epicentre', as the WHO did regarding corona in mid-March. Now there is a determinate reason for why Covid-19 selected members of dominant classes and inhabitants of advanced capitalist countries among its first victims, and to this we shall return. But the question here is another: what difference did it make?

Consider a counterfactual timeline of victimhood more similar to that of the climate crisis. Imagine that Covid-19 would have jumped from Iran into Iraq in February 2020, killing a couple of thousand in Basra and Baghdad, then leapfrogging to Haiti, killing another 5,000, before swerving down to Bolivia and Mozambique, taking out another few batches of the same size, while the number of patients hovered in the lower hundreds in

London, Paris and New York. It is not a far-fetched conjecture that governments of the global North would then have let the virus fester. They might have sent aid packages, perhaps offered some conditional debt relief, but they would not have put capitalism in quarantine, even if keeping it going as usual would have facilitated further transmission – and why should they? It would not have been *their* people dying, at least *not initially*. With the climate, the North has had decades to get used to the idea that while it might be stalked by extreme weather every now and then, and while the nuisance might become more severe, it is still, for the time being, *primarily* a problem on the other half of the planet, relayed by the steady background noise of the latest news of a disaster in some wretched periphery. With this come other priorities.

The ten countries with the most deaths from Covid-19 in the last days of March were, in descending order, the US, Italy, China, Spain, Germany, France, Iran, the UK, Switzerland and the Netherlands. As it happens, all of these countries except two (Iran and Switzerland) also made it to the top-ten list of territorial units responsible for the most cumulative CO_2 emissions since 1751. The countries that seemed worst afflicted by the pandemic in March 2020 were some of the very same countries – the US above all – that had caused most of the climate emergency: some way towards an explanation. This timeline of victimhood came as a shocker. One German liberal put it well:

In our imagination, it did not occur that intensive care units could burst at the seams and an adequate number of ventilators be lacking. Landing in the hallway of a crowded clinic with a high fever, out of breath, counted as a horror fantasy. So far, we have thought: this may be everyday life in developing countries, maybe also in Russia, but not among us!

And that changed, as it were, everything. The timeline of victimhood placed rich and poor at opposite ends for corona and climate: in the former case, inducing governments of the North to do the right thing; in the latter, to behave in a manner that can only be called evil. Perhaps humanity should thank Covid-19 for taking the early route through Europe.

But that can be only part of the explanation, for the varying temporalities also had implications for the perpetrators. When the parameters of the climate crisis were laid down in the late 1980s and early 1990s – this being the moment when the science matured, the UN directed its member states to reduce emissions, the issue settled in the Northern public consciousness and the first extreme weather events, notably the US heat wave of 1988, were linked to the trend – fossil capital launched a pre-emptive war. It spared no effort in sabotaging mitigation. Whether it stuck to literal denial or shifted to sundry smothering forms of green capitalism, it met every initiative for real action with resistance. Since then, measures to cut emissions in line with the need have been led astray, bogged down, repudiated, overhauled through myriad pathways, the focus lost, the quest coming to appear hopeless – yes, a labyrinth of traps and ambushes set up by the enemy.

By dint of being a secular trend, global heating gave *extended opportunities for obstruction* to the perpetrators, in conjunction with a poor-first timeline of victimhood. Covid-19 negated both. So suddenly did it strike that no capitalist interests had the time to build up apparatuses for resisting the suspension of business-as-usual. Transmission of the virus was perpetrated through travel, and so one could imagine that airlines, cruise operators, auto industries backed up by oil companies – incidentally, a fair share of fossil capital – should have tried to pre-empt or dilute the closures, just as they had done on the climate front. But the blitz of the virus overwhelmed even them. Instead they hunkered down in the state of exception, waiting for the pandemic to be defeated so things could return to normal. Granted, there were some feeble attempts from this corner to restore the priority of economy over health, but they were readily swept aside, at least in the initial phase – a lesson in politics as the art of the first blow or, if you will, war of manoeuvre.

Covid-19 came as an instantaneous and total saturation of everything. Like a gust blowing out the tinted windows in a skyscraper, it stripped the state down to its barest relative autonomy. Governments in the North were in a rare position to sacrifice the well-being of their capitalist economies for the lives of their elderly and potentially younger cohorts too. One may regard this moment as bringing out the best in modern bourgeois democracies, the respect for life trumping the respect for property, a victory for the egalitarian premise to which democracy is sworn (but then, tempering the congratulation, China and Iran acted earlier).

Whichever way we look, we are drawn back to the differences in time: global heating as secular, Covid-19 as shock. But this raises the possibility that the question is badly posed in the first place. Are the two at all comparable? Is this not a bit like juxtaposing the biography of one person with an hour in the life of another? We shall return to this problem shortly, but, for now, we shall instead add the dimension of space. Suppression of Covid-19 fits into the overarching paradigm that has taken hold of Northern politics in recent years: nationalism. It could be executed by closing borders, sending military to patrol them (Denmark jumped on this opportunity), promoting autarchy, shutting oneself off from the outside world. The benefits of such measures, insofar as they were effective, accrued straight to the national population. But when it comes to emissions cuts, gains would be distributed across the globe; Kenyans would benefit as much from deep cuts in Germany as the Germans and then some. Suppression of CO_2 does not sit naturally in the framework of the nation-state. The war against Covid-19 could be conceived as a classical war, drawing on all the paraphernalia of patriotic pride – a nation protecting itself, as in previous moments of danger; a people sheltering behind the bulwark of the state – whereas a war against CO_2 would tend to slip out of that mould. It would be a war for the benefit of one's own *and* foreign others. First of all, it would be a war for the poor.

Shifting shades of extremism

Corona and climate share one structural quality that invites comparison: the amount of death is a function of the amount of action or inaction on the part of states. Left untreated, both afflictions become self-amplifying – the more people infected, the more will be infected; the hotter the planet, the more feedback mechanisms heat it up further – and once underway, the sole way to terminate such spiralling burns is to cut the fuse. States in the global North have now offered proof that this is possible. It will not be easy to erase.

When climate activists, advocates and scientists demanded that emissions be cut, they were told that it would be too expensive: it might shave off one or two tenths of a per cent of the GDP. (The climate crisis 'does not justify policies that cost more than 0.1 percentage point of growth', the *Wall Street Journal* typically lectured in 2017.) Some might lose their jobs. There could be bankruptcies. People would never accept the disruptions to the lives they're used to, and anyway, if some were to cut their emissions, there would always be others – 'free riders' – wallowing in the pleasures of carbon.

As it turned out, all of this and much more went out the window in March 2020. No plan for a transition from fossil fuels ever involved the dislocation knowingly triggered that month: no one proposed that world capitalism should be paused overnight to save the climate. No one suggested emissions be cut by one quarter in thirty days – the demand for 5 or 10 per cent per year was brushed off as extremism beyond the pale. No one argued that humanity ought to be placed under curfew. No road map, no manifesto, no vision from the climate movement – and it has its share of radicals – ever sketched anything like the meteor storm of state interventions that hit the planet in March 2020, and yet we were always told that we were being unrealistic, unpragmatic, dreamers or alarmists. Never again should such lies be given a hearing.

Did the citizens of the world baulk at emergency action? A global poll conducted at the end of March indicated mass support for the measures, the discontent rather slated in the maximalist direction: nearly half of humanity, or 43 per cent, thought their governments were doing *too little* to fight the pandemic. Only eight of the forty-five polled countries registered significant minorities – around a quarter of the populations – unhappy with what they considered their governments' overreaction (this included the US, homeland of do-nothing-on-climate, where the share stood at 19 per cent). Apparently, the idea that some things will have to stand aside when many lives are at stake wasn't too difficult to sell to the public.

Did free riders subvert the measures? Corona and climate could both be framed as the 'collective action problems' beloved by game theorists: all would enjoy the fruits of cooperation – a vanquished pandemic, a stabilised climate – but any one individual could slip away and go on vacation in a hot spot, refrain from washing his hands, speak five centimetres from the face of his interlocutor or indulge in extravagant emissions and be doubly better off for it. He would not have to give up any dear habit while taking the same advantage of the aggregate efforts as anybody else. But if everyone were to act that way, all would be in vain – the collective action would unravel, as those hesitant about climate mitigation have foreseen. In the event of corona, however, states cut through this Gordian knot by simply *laying down rules* for citizens to abide by. They could not eliminate all shirking and free riding, nor did the guilt-tripping and the repression come without their own set of problems, but on the whole, given the far-reaching nature of the demands put on people, the collective action launched from above was remarkable for its cohesion. Clearly the dilemma is not irresolvable.

And then one should remember, again, that no champion of radical emissions cuts has ever asked people to submit to something as unpleasant as a lockdown. Climate mitigation would never require people to become hermits in their homes. In fact, convivial living would be conducive to that project: riding a bus, sharing a meal, having a daggering party in the street, spending time with loved ones in retirement homes or paying for a concert instead of the latest console from Amazon would be *in line* with the endeavour to live *sans* fossil fuels. Not only could a climate emergency programme skip the interferences with rudimentary mobility in space, it could offer *improvements* in the quality of life, as the movement has demonstrated by years of propaganda and praxis on subnational scales. Such win-wins merely presuppose enemy losses.

When the coronavirus that caused Covid-19 advanced through the world in early 2020, it was not, in and of itself, an effluent from profit. It did not emanate directly from the chimneys of accumulation. CO_2, on the other hand, is the exhaust gas from the material substratum of surplus-value production – fossil fuels – and thereby a coefficient of power. There are interests at stake in its continued release into the atmosphere. Hence parts of fossil capital and its friends have spent decades preaching that higher concentrations of CO_2 are actually *good* for humanity, but no one has yet made such a plea on behalf of the coronavirus. Another level of antagonism suffuses the climate. An enemy of higher order must be overcome, and not for a month or a year or two: the shutdown of fossil capital would have to be permanent. The emergency itself would, of course, not be everlasting; there would be a transitional period, with effects outlasting it, or else it would have failed. The period might be longer than for Covid-19 – although, as of this writing, no one can tell how long this will go on – but it should be considerably less painful. It would entail a more thoroughgoing breakdown of private property. It would bury forms of capital for good. It would be something more akin to war communism.

2

Chronic Emergency

On a closer look, however, the appearance of energetic action against the pandemic is but a semblance. The contrast between coronavirus vigilance and climate complacency is illusory. The writing has been on the wall about zoonotic spillover for years, and states have done as much to address it as they have done to tackle anthropogenic climate change: nothing. 'Zoonotic spillover' is less of a household term – but that should now change – referring to an infection that first sits in an animal and then jumps into a human. A pathogen spills over the species boundary. It could be a worm, a fungus, a bacterium, an amoeba or a virus; of whatever sort, the pathogen is a miniscule creature that eats its prey from within. Paragon of the parasitic, it infiltrates a body and leads its existence inside it, feeding, reproducing and, in the process, inflicting damage upon the host.

'Coronavirus' is a family of viruses with special proficiency in this regard. It gets its name from the appearance of the molecule under the microscope: a greyish ball with dozens of red spikes, looking somewhat like a royal crown or *corona* in Latin – a ubiquitous image in the spring of 2020, crowning it the organism if not person of the year. With the spikes working like hooks, the virus can drive itself into other cells and hold on to them. Like so many others in its family, this particular coronavirus, formally designated SARS-CoV-2 by the WHO, escaped from its original hosts among bats. But why would it ever do that?

Under normal conditions, coronaviruses and other zoonotic pathogens lead an inconspicuous existence in the wild. They hitch ride after ride on their natural or 'reservoir' hosts – an animal that harbours the parasite and puts up with it, suffering little if any illness. Over millions of years, the viruses have co-evolved and reached a *modus vivendi* with these hosts, permanently inhabiting their bodies without killing them. Sometimes a couple of monkeys or mice might fall ill and drop dead on the forest floor, but the generous vegetation would scoop up their carcasses before humans had reason to notice.

Tropical forests house the greatest abundance of species; their ranks thin out near the poles. Ice ages have periodically wiped the slates of evolution clean on high latitudes, where there is less insolation. Around the equator, flora and fauna have been spared glaciers and luxuriated in the energy streaming in from the sun, making tropical forests nurseries of the most astounding biotic exuberance. They also have the richest pathogen pools. The closer to the equator, the more hosts and invisible riders, some of which may on occasion strike out into new terrain. For them to succeed, a number of conditions must be fulfilled: the reservoir hosts must shed the pathogen – as in sneezing or coughing or bleeding it out – onto another host, which must be susceptible to infection. If the pathogen is lucky, it happens to be an 'amplifier host', in whose bosom the agent of disease can multiply profusely, try out new genetic combinations, gain momentum and prepare for the next step, which must be similarly successful. Most links of transmission soon reach a dead end. Every now and then, however, openings in the barriers align and the pathogens make it all the way into human populations. The shorter the distance, the less of a feat it will be.

This is an old story: bubonic plague and rabies are two notorious examples of zoonotic spillover. They don't seem like particularly modern distempers, at home among flush toilets that smell of perfume, which is why the problem was fairly recently consigned to the past. In the decades after World War II, the most golden age of capitalism, one could learn that 'the Western world has virtually eliminated death due to infectious disease'. Such Panglossian diagnoses somehow carried over into the last years of the second decade of the millennium. In his airport bestseller from 2018, *Enlightenment Now*, Steven Pinker, the leading voice in the choir of bourgeois optimism, revelled in the 'conquest of infectious disease' all over the globe – Europe, America, but above all the developing countries – as proof that 'a rich world is a healthier world', or, in transparent terms, that a world under the thumb of capital is the best of all possible worlds. ' "Smallpox was an infectious disease" ', Pinker read on Wikipedia – 'yes, "smallpox *was*" '; it exists no more, and the diseases not yet obliterated are being rapidly decimated. Pinker closed the book on the subject by confidently predicting that no pandemic would strike the world in the foreseeable future. Had he cared to read the science, he would have known that waves from a rising tide were already crashing against the fortress he so dearly wished to defend.

He could, for instance, have opened the pages of *Nature*, where a team of scientists in 2008 analysed 335 outbreaks of 'emerging infectious diseases' since 1940 and found that their number had 'risen significantly over time'. Most were cases of zoonotic spillover, the lion's share of which originated in the wild. A survey published six years later observed the same trend, but identified a gearshift in the 1980s, the decade of HIV, the most renowned modern virus to spill over from animals before SARS-CoV-2 came along. Since then, the list of pathogens imported from other species has extended like a record of running transactions: the Nipah virus, first detected in 1998 in Malaysia; the West Nile virus, coming to New York in 1999; Ebola, striking West Africa to devastating effect in 2014; Zika, rolling through Latin America and the Caribbean in 2015; the coronavirus that caused SARS, rattling the world in 2002; the coronavirus that caused MERS, making the rounds in the Middle East in 2012; a slew of old diseases staging comebacks, sometimes with novel strains – anthrax, Lyme, Lassa fever – and a series of influenzas appearing with the regularity of hurricanes, but given more faceless names: H1N1, H1N2v, H3N2v, H5N1, H5N2, H5Nx and so on. By 2019, the scientific literature referred habitually to the fact that 'infectious diseases are emerging globally at an unprecedented rate', the share made up of zoonoses estimated at between two thirds and three fourths, increasing to nearly 100 per cent for pandemics. It is a secular trend in its own right.

That strange new diseases should emerge from the wild is, in a manner of speaking, logical: beyond human dominion is where unknown pathogens reside. But that realm could be left in some peace. If it weren't for the economy operated by humans constantly assailing the wild, encroaching upon it, tearing into it, chopping it up, destroying it with a zeal bordering on lust for extermination, these things wouldn't happen. The pathogens would not come leaping towards us. They would be secure among their natural hosts. But when those hosts are cornered, stressed, expelled and killed, they have two options: go extinct or jump. In his now must-read *Spillover: Animal Infections and the Next Human Pandemic*, published in 2012, David Quammen likens the effect to the demolition of a warehouse. 'When the trees fall and the native animals are slaughtered, the native germs fly like dust' from under the bulldozers. The science is agreed: the secular trend has a very general driver in the economy advancing from the human side all over the wild. Another turn to the non-human world is, after this, a must. It begins with the order Chiroptera.

Of bats and capitalists

The world is home to upwards of 1,200 species of bats, as of 2020. With speciation ongoing for at least 65 million years, this is one of the oldest orders of mammals, the second most diverse – only rodents exhibit greater variety – accounting for one fifth of all extant mammalian species. Bats are also the nonpareil carrier of pathogens. While rodents are likely to carry slightly more viruses in absolute terms, on account of their multitudes, bats host far more such guests per species – and yet they do not seem to mind. They are persistently infected, without sign of malaise. There are no reports of mass die-offs of sick colonies. Hence chiropterologists have postulated that bats possess a unique tolerance of viruses, exceptionally powerful immune systems that must spring from some common trait conferred by those millions of years of evolution. What could it be?

All bats have one hallmark ability: they can fly. While some squirrels and lemurs glide or parachute short distances, bats are the sole mammals to power sustained flight by frenetically flapping their wings. This activity does not come for free. To stay in the air, bats have to expend prodigious amounts of energy, driving metabolic rates to the point where their bodily temperatures reach 40°C – think marathon runners – for hours on end. Less sprightly mammals would experience this condition as fever. Fever, of course, is a primary defence mechanism for bodies beset by illness, which seems to imply that bats

– for whom this is more like a natural state – could easily produce a little fever to shrug one off. Conversely, viruses that settle on bats must adapt to feverlike temperatures. The theory, then, is that bats have become bearers of pathogens that cannot impair *their* constitution, but can overpower the weaker immune systems of other mammals. Flight has other consequences too: it allows bats to travel over vast distances – dozens of kilometres each night in search of food, hundreds between roosting sites, more than one thousand between summer and winter grounds – and pick up and disperse pathogens along the way. Bats spend little time on the ground and much in the air, in trees, under roofs, in positions from which they can let drop saliva and excrements on whatever is beneath them. They can move close to humans, into their orchards, fields, houses and stables, if they have reason to.

And bats have a second key trait: they are gregarious. They huddle together in clusters of uncommon density and diversity. Some bats squeeze in 3,000 individuals per square metre and several million per roost; some hang out in ensembles of multiple species, swapping viruses back and forth – a paradise for pathogens and their evolution, and an ideal formula for herd immunity. Bats, in other words, live by breaking the two principal rules of the 2020 lockdowns: do not travel and do not form crowds. This would explain why they are such hypercompetent reservoir hosts, and why their viruses can become so virulent in other settings, as the world has had repeated occasion to learn.

The Nipah virus spilled over to humans in a forested area in northern Malaysia when bats were attracted to fruit trees around a pig farm. They shat or otherwise excreted onto the pigs, which served as amplifier hosts, passing the virus on to humans; it caused fever, cough and shortness of breath, worsening into confusion, coma and inflammation of the brain, with a mortality rate nearing 40 per cent. Some 110 people died in the initial outbreak before it was contained. Rabies is deposited in bat reservoirs. So are dozens of other pathogens of well-documented malignancy, including, probably, Ebola.

But the prime speciality of the Chiroptera is corona. SARS was the first coronavirus to unleash a pandemic in the new millennium, taking scientists by surprise and sending them into the caves of southern China to identify horseshoe bats as the reservoir host, whence the virus switched to civets as the amplifier before seizing on humans. The discovery of corona in bats dates to the early years of the millennium, and barely had it been made before the virus struck again: MERS leapt from bats to camels to humans. After these brushes with mass death, more scientists fanned out across the tropics to try to get an idea of what was in store. One team affiliated with the aptly named PREDICT project – the largest of its kind – captured some 12,000 bats in twenty tropical countries, collected swab samples and returned the animals to the wild and found that the number of coronaviruses per species came close to three. (Thousands of rodents and primates were also tested, but more than 98 per cent of the positive individuals were bats.) This led them to put the probable number of distinct coronaviruses circulating in the whole planet of bats at 3,000. Far from all would be capable of infecting humans – in 2016, a bat colony in southern China dropped a coronavirus onto pigs that made them die from acute diarrhoea, but it failed to penetrate people – although several hundred are likely to have that potential, and many more might be on the way.

Once it hijacks a cell, a coronavirus behaves like a living creature, and indeed coronaviruses are subject to natural selection. They strive to adapt to their surroundings – batten onto the host, survive attacks, exit to another, replicate, perpetuate the lineage – and endure only insofar as they manage these tasks well. And coronaviruses can evolve faster than most. Their genetic information is encoded not in the complex double helix of the DNA, but in the simpler RNA, a molecule with a single strand that can mutate with fantastic velocity – think a relay of sprinters outpacing a heavy carriage – throwing up new genetic combinations to try out against the environment.

SARS-CoV-2 had struck on one particularly splendid advantage: it could leap from host to host *before* inflicting the damage. One human passed it on to the next prior to developing symptoms. Hence the chain of transmission stretched out over continents far more effectively than for SARS, which had the reverse profile – first symptoms, then peak infectivity – and was broken with relative ease. Once again the virus stemmed from bats, and more strains of coronavirus are, for a certainty, being hatched. So why don't we just kill them all? The outbreak of Covid-19 in China prompted calls for bat populations to be eradicated. This is a common response to spillovers – Nipah, to take a minor example, induced the slaughter of more than one million pigs near the source – and appears to make some sense. It could be extrapolated further. Why don't we pave over what remains of the wild? If the whole planet looked like Manhattan, surely there wouldn't be so many parasites to pester us.

Insane as the idea might sound, there is an element of logic to it. General biodiversity should correlate with diversity of pathogens. If the former is slashed, one might expect that whole clades of reservoirs, amplifiers and parasites would be taken out. But it can also go the other way. Biologists have put forth the hypothesis of a 'dilution effect', according to which a richness of species *ipso facto* inhibits spillover. If an abundance of animals is present in an ecosystem – say, a forest – some will be incompetent hosts, in which parasites will find scant nourishment and platforms for replication; when biting them, the attempted transmissions go to waste. If, for example, there are plenty of squirrels in a forest, they will take some of the bites from ticks that might otherwise target humans. The unpleasant, sometimes debilitating Lyme disease – in the worst case causing chronic fatigue and cognitive disorders – is conveyed by ticks, which have been shown to waste many of their bites on a species of opossum present in biodiverse forests of North America. The opossum kills the ticks. But in degraded forests it vanishes, while the white-footed mouse, a most competent and tolerant host for the ticks, continues to thrive – indeed more so than ever, as it is relieved of competitors. The depletion of biodiversity removes the buffers.

While the dilution effect remains the subject of theoretical controversy, there is now across-the-board empirical evidence for it as a law: *higher biodiversity means lower risk for zoonotic spillover.* The species that survive assaults on wild habitats tend to be the opportunists and generalists – think mice or weeds – that carry pathogens with ease, reproduce at speed and make themselves at home in the interstices of human settlements. The warehouse can be demolished, but the dust will not go away. In the process of demolition, it will blow straight in our direction.

It wouldn't be the first time humans reacted to infection by taking it out on the bats. In Australia, the flying fox has been harassed and hunted by upset grim men ('They shit on people! It's backwards – let the people shit on them!' Quammen quotes one supporter of the chase. 'What good are they? Get rid of them! Why doesn't that happen? Because the sentimental greenies won't have it!') In Brazil, roosts of vampire bats have been systematically blown up with explosives. But in all investigated cases of culling, the result has been the opposite of the intended: the pathogen loads have been scattered farther afield. Eradicating bats would merely be one more way to lose biodiversity, as they perform critical functions in pollinating plants, dispersing their seeds and keeping pest insects in check. So far, however, the desire to avenge pandemics is a negligible threat. Instead we must examine deforestation.

Deforestation is an engine not only of biodiversity loss, but of zoonotic spillover itself. When roads are cut through tropical forests, patches cleared, outposts placed deeper in the interior, humans come in contact with all the teeming life forms hitherto left on their own. People raid or occupy spaces where pathogens dwell in the greatest plenitude. The two parties stage their most frequent encounters along the edges of fragmented forests, where the contents of the woods can slip out and meet the extremities of the human economy; and, as it happens, generalists like mice and mosquitos, with a knack for serving as 'bridge hosts', tend to flourish in those zones.

Fragmentation is now the fate of the planet's forests. Some 20 per cent of remaining forested area stands within 100 metres of an edge and some 70 per cent within one kilometre: archipelagos of wooded islands in oceans of cleared landscapes. On the whole, this is a bane for biodiversity, but again less so for parasites. One group of ecologists has recently advanced the intriguing hypothesis that fragmentation *accelerates* the evolution of pathogens, by locking them and their hosts into island-like habitats and pressing them to come up with paths to survival in the restricted space. On each island, there would now be a 'coevolutionary engine' of parasite and host, taking the most advantage of any mutation and genetic drift and driving down its own separate trajectory, so that, paradoxically, pathogen diversity is increased. The engine of deforestation acts to rev up spin-off engines of parasitic experimentation. And this would be going on right next door to the interface with humans.

Whether that particular hypothesis is confirmed or not, it is evident that the hotspots of spillover are the hotspots of deforestation: and they are located in the tropics. That's where the greatest abundance of bats is found. A quarter of the world's bat fauna lives in Southeast Asia, where chainsaws and bulldozers have been crashing through tropical forests in recent decades. The result seems to be imposition of chronic stress. Bats have to make up for lost shelter and food, fly back and forth between island patches, navigate hazards and cross enemy territory – a more stressful life than in the contiguous forests of old. How does it impact their health? Much like stress wears out human bodies. In Sabah, a Malaysian province on the northern tip of Borneo, one team of chiropterologists placed traps in and around forest fragments and examined the bodies of the captured bats. They turned out to have smaller body mass, fewer white blood cells – the infantry of the immune system – and generally poorer constitutions than their conspecifics in less disturbed areas. The stress caused by deforestation appears to crack the otherwise impervious defences of bats and trigger 'pulses of viral excretion' – episodes when viruses are shed en masse onto accidental hosts, who might well be humans, as bats deprived of their old habitats seek shelter and food in barns, gardens, villages and plantations. (Bats can do rather well on cacao plantations.) After the forests have fallen in eastern Australia, the flying fox has little choice but to subsist on what the ranches and parks offer. When Brazilian rainforests are cleared for pastures, the vampire bat, so called because it feeds on blood, is pushed to attack cattle. Similar dynamics come into play for rodents.

Now, if deforestation drives zoonotic spillover in the early twenty-first century, we must ask: what drives deforestation? The cutting down of trees is, of course, an ancient human practice. Like any other such practice, it takes on shifting forms over time. A break occurred in the 1990s. Before that decade, deforestation in the tropics – particularly Southeast Asia and Latin America – was largely initiated by the state. In the 1960s and '70s, peasant insurgencies billowed through the continents, operating from within remote forests in every newly independent country in Southeast Asia – one, two, many Vietnams – and seeking to bring the example of the Cuban Revolution to most corners of Latin America. The US enjoined the governments under its tutelage to stem the tide by colonising their hinterlands, so as to strip the insurgents of cover and undercut their popular support. If smallholders were given the land they craved, they would not go over to the guerrillas, and far better than to expropriate the estates – the very thing the US sought to avoid – was to open up the forests. Hence the military dictatorship of Indonesia resettled smallholders on the outer islands, while that in Brazil bisected the Amazon with a mega-highway and dispatched pioneers to stake out their own land claims along feeder roads. Come the 1990s and the insurgencies had all been defeated, while the structural adjustment programmes compelled states to improve their finances and trade balances. The drivers were inverted. A global shift in the balance of class forces caused a transmutation in deforestation: it became 'enterprise-driven', in the terms of a seminal meta-analysis. The state took a step back and concentrated on transferring titles to cheap land and labour. The initiatives to clear forests now came from 'highly capitalized, well-organized' private actors, who built the roads and sent their machines to make way for plantations, quarries and ranches or to log the timber, with an eye to some distant market. The state no longer advanced into the forests with a troop of ragged smallholders behind it: another force smashed into them for its own purposes.

In the new millennium, it is the production of commodities that chews up tropical forests. It negates diversity on every front. No more than four commodities – beef, soybean, palm oil and wood products, in descending order of impact – accounted for four tenths of the dramatically sped-up tropical deforestation between 2000 and 2011, split among seven countries in Southeast Asia and Latin America. The historical break was visible from above. Smallholders make for small clearings, while enterprises with operations on industrial scale make for large ones, palm oil plantations regularly covering more than 3,000 hectares in Indonesia and cattle ranches more than 1,000 in Brazil. Hence the size of a clearing is a proxy of the driver, and the most recent available analyses of satellite maps show that the bulk of deforestation in the first decade of the millennium took the shape of large and medium-size clearings, beyond the means of smallholders; the trend was most salient, again, in Southeast Asia and Latin America, scenes of the largest total losses.

What stresses the bats in Sabah? It is palm oil plantations that creep up on their habitats. Malaysia and Indonesia together produce 90 per cent of all palm oil in the world, 70 per cent of agricultural land in the former country now devoted to this one commodity. It puts Malaysia in the top league for the largest clearings. No other nation lost old-growth forests as fast during the first decade of the century, when they were razed to make room for mono-crop fields interspersed with mills; the oil must be squeezed from the palm within twenty-four hours of harvest. The plantations are owned by some of Malaysia's biggest companies, fully integrated in global financial circuits, indispensable as a source of upfront investments, the mills and refineries plugged into networks of roads, trucks, harbours and tankers that can ferry the product to any market. Thousands of workers lodge on the plantations in huts provided by the companies. In Sabah, one researcher uncovered a pattern of predominantly immigrant workers held in debt bondage – indebted upon arrival; passports confiscated; intimidated by police; food available only on credit, compounding their debt – and often paid below the legal minimum wage. What do these enterprises aim to achieve? 'Plantation companies and their shareholders', writes one group of forest ecologists, 'seek to maximize the marginal returns to their capital and seek access to large expanses of cheap, unencumbered land with access to reliable low-cost labour.' These are – there is no other term for it – eminently capitalist enterprises. As such they are subject to the compulsion to expand. Southeast Asia has seen an epic resurgence of the plantation in the early twenty-first century, a unit from the colonial era coming back to grab land and squeeze out other forms of life. The integrity of bats would be the least of the owners' concerns.

Hence the region is the theatre of an undeclared war on flying foxes, the largest megabat with a wingspan of nearly two metres, normally congregating in noisy communal camps in mangroves, swamps and rainforests. Deforestation rips up their habitat and stresses them out. Palm oil is not the sole commodity to blame, of course; in the case of the Nipah virus, it was commercial pig farming that drove wedges deep into the woods of fruit bats and forced them to visit the farms. Deforestation-induced stress has been reported for a bat species in a northern area of Thailand likewise beset by plantations and infrastructure development, although not quite on the scale of Malaysia. The number of high-resolution studies remains low at the time of this writing, but chiropterologists recognise the broad picture, and it is not confined to Southeast Asia.

The Cockpit Country in the heart of Jamaica is a karst landscape, formed through the dissolution of limestone rocks, taking the shape of many hundreds of mountain peaks around deep valleys that continue underground in uncountable caves. As rich in shelter, the forest is in food: wild yam and banana and guava trees, cotton, Santa Maria, guango. This is a haven for more than a dozen endemic bat species. During the colonial era, it was also a refuge for slaves on the run from plantations, or maroons, who fought the British from within the nearly impenetrable fastness. According to oral traditions, the maroons would replenish their muskets with gunpowder concocted out of bat guano from the caves, so rich in nitrogen as to be explosive. Maroon communities descending from the runaway slaves still inhabit the villages around the forest and act as its custodians, but they face a threat: there is bauxite in the ground. The Canadian mining company Noranda has been chafing at the bit for years, waiting for the go-ahead to enter and commence strip mining, but the government has so far hesitated. The fate of Cockpit Country is the central environmental battle of Jamaica. It attracts a steady stream of protests from maroons and their allies and counter-protests from Noranda's employees and is invoked every time there is a climate rally, as during the school strikes of 2019. When the Covid-19 pandemic reached Jamaica, calls for the elimination of bats prompted one of the environmental NGOs most involved in the struggle to instead call out: 'Leave the bats alone!' – another reason not to unfetter Noranda. How many more interfaces of this kind are in the offing? Nobody knows. The points of collision between bats and capitalists around the equator remain to be mapped.

But one scientist who has begun the work is Rob Wallace, heir apparent of the venerable tradition of dialectical biology in the age of zoonotic spillover. He has honed in on Ebola. This virus, of another family than corona, had been slumbering in West Africa for a long time, riding on fruit bats, skipping out of the rainforest to infect a village or two, the outbreaks documented since 1976 coming to two dozen. What happened in 2014 was something altogether different. In an outbreak more than forty times larger than any before it, Ebola vaulted into a proto-pandemic phase, striking out from Guinea to hit Liberia, Sierra Leone, Nigeria and Senegal, killing more than half of infected patients – fever, diarrhoea, profuse bleeding inside and out of the body – and leaving corpses on the streets of capital cities. What accounted for this phase change? It must be some 'non-virological' development, not internal to the pathogen or its host. In the half-decade before the disaster, the World Bank had identified the forested zones of the region as 'one of the largest under-used agricultural reserves in the world', and the government of Guinea set to work handing it over to one branch of agribusiness: palm oil. The plantation boom arrived here too, promoted by companies from countries such as the US, the UK, Malaysia and Indonesia, expropriating swaths of land for the commodity. And naturally, fruit bats were kicked out of old haunts and instead swarmed around the palm groves. By denuding old-growth forests, the enterprises did away with the 'friction, which typically keeps the virus from lining up enough transmission', the dilution effect inverted into streamlining. None of the natural scientists we have referred to up to this point are Marxists (although one cannot, of course, rule out that they are closet Marxists), but it takes a card-carrying Marxist like Wallace to draw out the implications: 'opening the forests to global circuits of capital' is in itself '*a primary cause*' of all this sickness. It is unrestrained capital accumulation that so violently shakes the tree where bats and other animals live. Out falls a drizzle of viruses.

Ecologically unequal and pathological exchange

There is another implication too: causation is not local. 'If landscapes, and by extension their associated pathogens, are globalized by circuits of capital, the source of a disease may be more than merely the country in which the pathogen first appeared.' The impetus for mining even more bauxite in Jamaica comes from elsewhere. The material is not taken up to feed the children of Kingston. It is foreign capital investing in it, shipping it off to aluminium plants in the US and profiting from it, in the very picture of scorched-earth extractivism, leaving behind gaping red wounds in the landscape and asthma for the kids. The palm oil is not meant for the households of Malaysia. It is exported to cosmetic, chemical, food, livestock and energy industries around the world, and the trend is the same for the other three commodities felling the tropics: beef, soybean and timber flow out of their countries of origin. To make sense of this, scholars have borrowed the concept of 'teleconnections' from meteorology – a rise in the atmospheric pressure in the Malay Archipelago sets off an El Niño that touches off downpours over Peru and dry spells in Botswana – and applied it to trade flows. Remote demand for products grown in the tropics now drives deforestation in the aggregate.

This should not come as a surprise, as late capitalist globalisation is defined by spatial separation between production and consumption: what is bought in one place comes from some unfamiliar antipode. Causation must then levitate above the ground, but it doesn't move evenly. The atmospheric pressure of demand still rises in the North. If one calculates the amount of land embodied in traded goods – the land required to grow the commodities, feed them, mine them, process and assemble them – one finds Europe to be an epicentre of teleconnections. EU countries source more than half of their land-based consumption from other parts of the world, the share exceeding four fifths for Germany. Japan (92 per cent) and the United States (33 per cent) are not far behind in total appropriation of embodied land. These are not trivial flows. Just how much land is shuffled northwards is indicated by the calculation that in 2007, the EU had a net import of goods embodying land *as large as the entire surface area of India*; it will have grown since then. Put differently, every year the EU sucks in a volume of land at least equalling the size of India, over and above the land embodied in its export and the import that matches it. The trade balance might show something else; it can record a surplus of exports or a small deficit, as counted in euros or dollars. But counted in *actual land*, the EU gobbles up enormous quantities offered up by others through the mediation of the market, with no balance in sight. The Union epitomises the process known as ecologically unequal exchange: transactions that might seem fair on the monetary surface, but allow rich countries to absorb biophysical resources from the poor and drain their natural endowments.

These include forests. With all the rigour of quantitative methods, a body of literature has demonstrated that if developing countries rely on export to developed partners, they tend to cut down their own forests faster to serve up the commodities in demand. Such relations would not, of course, have come about if it weren't for colonialism. The legacy is kept alive by means including structural adjustment programmes, debt repayment, investment from multinational corporations and projects supported by state agencies such as the US Export-Import Bank, all of which have been shown to speed up forest loss. Out of the clearings the commodities can then come gushing: the American appetite for hamburger is satisfied from pastures carved out of the Amazon. The import of coffee to the North presupposes deforestation in the tropical belt. Chocolate, consumed in the most tremendous quantities in Switzerland, Germany and Austria and supplied by a mirroring top trio of Ivory Coast, Ghana and Indonesia, comes from cocoa trees grown where wild forests once stood: and the shopping list goes on.

Far from the shelves of supermarkets, on latitudes closer to the equator, this translates into the ravaging of local biodiversity. By purchasing commodities from the tropics, rich importers can offload the impacts on animals and plants from their own lands to those that, incidentally, house a greater richness of species. It has taken the science some time to catch up and connect the dots, but, in 2012, one pathbreaking study derived one third of all existential threats to animal species straight from the sale of goods like coffee, beef, tea, sugar and palm oil to countries of the North. The top seven importers of biodiversity threats were – always on top? – the United States, Japan, Germany, France, UK, Italy and Spain. (When these words are written, in mid-April 2020, the top seven countries with the most registered cases of Covid-19 are the US, Spain, Italy, France, Germany, the UK and China. We shall have to look at the meaning of this coincidence below.) The top seven exporters of these same threats – that is, countries from which biodiversity was bled white – were Indonesia, Madagascar, Papua New Guinea, Malaysia, the Philippines, Sri Lanka and Thailand. The flow is vertical, the space of species blowing to the North.

A suite of studies confirming the basic pattern has followed, some putting the share of extinction threats to northbound exports much higher – up to 60 per cent – others staying closer to one third. One critical nuance has been added. Measured per capita, the variations in consumptive claims on biodiversity are even more skewed, with rich but sparsely populated countries like Canada and Finland shooting up to the top: one typical Finn will inflict species losses through trade far above the global average. One Chinese or Indian still treads lightly. The next step, of course, would be to break these numbers down further into income brackets and study how *classes* of Canadians and Finns, Chinese and Indians exert their respective pressure; but the findings can be pre-empted virtually by tautology. To be very rich means to have all the money in the world to eat tropical land. Indeed, statistically, 'with an increase in affluence biodiversity losses due to import increase faster than domestic losses' – that is, the richer one is, the more likely to gobble up the space of other species from a distance. Any bad taste is unlikely to survive the stage of packaging.

While ecologically unequal exchange has been traditionally arraigned as unethical for causing undue harm to the lands of the periphery, we can now, following Wallace's lead, take one step further: it is a deep driver of deforestation, hence of biodiversity loss, hence of zoonotic spillover. Some bats and other hosts will be sucked into those trade winds. The research is still in its infancy, as we have seen, but, in early March 2020, *Nature Communications* published a model study that followed the link all the way from shelf to sickbed in one case: malaria, one of those beneath-the-radar diseases, affecting some 230 million and killing 400,000 per year, the vast majority in rainforest biomes. Deforestation is a boost for the mosquito vectors. More sunlight reaches the soil where the larvae develop; when biodiversity retreats, fewer animals prey on

them. Nigeria suffers most from malaria due to deforestation. It is largely caused by the export of timber and cocoa.
Such commodities end up in the north: the consumers
with the greatest malaria footprint are the cocoa-guzzling
Dutch and Belgians, Swiss and Germans. 'In this unequal
value chain, ecosystem degradation and malaria risk are
borne by low-income producers' – or, in plainer terms: the
Europeans get the chocolate and the profits, the Africans
the mosquitos.

Bat-borne viruses travel farther with a greater capacity
for blowback, but the drivers are broadly similar. Indeed,
if one adds investment flows, one ends up with an even
more pronounced version of Wallace's upside-down map:
the real hot spots of disease lie in places like New York,
London and Hong Kong. The forces reaching out to
forests and pulling out pathogens are nowhere as strong as
in the central nodes of capital.

From the bush to billions of dollars

But it would be vulgar and ridiculous to argue that *everything* that happens in poor countries is the fault of the rich.
What about, for example, their own population growth?
One of the most comprehensive studies to date did identify
a 'mildly significant' correlation between total population
growth and deforestation. Another judged it 'non-
significant', while others have discerned a clear positive
association between the two. The results, in short, are
mixed, and sometimes counterintuitive. Rural population

growth, such as that incited by states colonising their forests before the 1990s, would seem to be the most problematic, but the reverse flows – people moving from forested areas into towns – might actually be more detrimental to biodiversity, since town-dwellers add their purchasing power to the suction of urban markets. They tend, in particular, to eat more meat. While the towns sprawl into nearby forests, elongating the periurban interfaces with wildlife, residents will go to markets to buy food taken from the back country. Some of it might be bushmeat.

'Bushmeat' is a term for the meat of animals hunted in the wild for the purpose of eating them, but elk shot by Swedish men and dragged back to their cosy fires do not usually fall under the heading. It is de facto reserved for the tropics. As such, killing a mammal on the savannah or inside the jungle is a practice even more ancient than cutting down trees, but it has now become an integral component of the sixth mass extinction. As of 2016, 301 terrestrial mammal species were being pushed towards the brink by hunters out for their meat, more than a third of them in Southeast Asia. The inventory includes Chiroptera, the order with the third most species endangered by this appetite; from flying foxes in Indonesia to fruit bats in Equatorial Guinea, bats are eaten on a daily basis, once again with Southeast Asia as the chophouse *primus inter pares*. Rodentia is amply represented too. But most of the species thus threatened are primates, the third mammalian order with a singular capacity to host pathogens; if bats have superior hospitality and rodents unmatched

tenacity, primates are of interest because of their phylogenetic proximity to humans. What they have, we can easily
get, and vice versa.

The zoonotic riskiness of this business is then readily
apparent. Indeed, hunters of bushmeat are frequently
infected by viruses, particularly from monkeys and apes,
although few make the crucial next leap from person to
person; there is a lot of 'viral chatter' and little of the real
thing. Of the infectious diseases spilled over between 1940
and 2005, a mere 2 per cent could be imputed to bushmeat, as against 44 per cent to 'land-use changes', 'food
industry changes' and 'agricultural intensification' – with
deforestation the pivot – but it shouldn't therefore be
discounted. It can be grave for plenty of forms of life. In
the teak forests of Myanmar, hunters track down tigers,
leopards, langurs, bears, tortoises and other animals on
the Red List of endangered species; in Central Africa,
mandrills and gorillas are beleaguered by snares and traps,
bows and arrows. Many hunters capture bushmeat for its
use-value, with no exchange-value involved: they bring it
back to their own households for cooking or medicinal
use. On the lowest rungs of global class society, they can
be protein-starved subsistence farmers, day labourers,
dispossessed peasants, members of ethnic minorities or
indigenous peoples, which makes this vehicle of extinction vaguely embarrassing for Marxists, best swept under
the carpet of more well-worn motifs. Bushmeat might be
a delicate problem. But to pretend it doesn't exist would
be outright dishonest.

It wouldn't be such a problem, however, if it hadn't been hitched onto powerful tendencies in the present. Deforestation prises open the basins of wildlife. Roads are so many launching pads for incursions; from the road-sides, hunters can place strings of traps within easy reach or go for a raid with guns. They can then drive off with the carcasses along roads that connect to urban markets. Bushmeat becomes a commodity to sell, age-old use-value no longer the aim of the hunt, which instead falls under the spell of profit. The first study of a commodity chain for great ape meat in Africa – chimpanzees and gorillas poached from a reserve in Cameroon, butchered into pieces, smoked and sold – found that 'hunters, all men, were primarily driven by profit, encouraged by middle-men' who pocketed the largest gains in the chain, as they made scant outlays. The nature of demand is consequently altered. Some bushmeat is upgraded into a delicacy served to affluent diners: bats as luxury dishes in Malaysia, lemurs in Madagascar. There occurs a qualitative leap into a global predicament described in a timely survey in *Science*, published in October 2019 and opening with a sentence pregnant with meaning: 'Wildlife trade is a multibillion-dollar industry that is driving species toward extinction.' It has become a department of accumulation of its own, trading not indirectly in the death of animals, like the businesses that grow from deforestation, but directly in their carcasses as luxury food and medicinal ingredients, ornamentation and clothing, or condemning them to a life as pets. With an annual revenue stream somewhere between 8 billion and 21 billion US dollars, this is now one of the largest of the many shady sectors of world trade, trawling through what remains of wild nature. The *Science* survey estimated that 5,579 species are dragged into commodity chains. That amounts to 18 per cent of all birds, mammals, amphibians and reptiles on the planet; but if the industry continues to expand – as it is prone to – the figure will soon rise to a projected 8,775 species 'at risk of extinction from trade'. And as this anti-ark is submerged in the oceans of money, some pathogens will look for other hosts.

Standard models in neoclassical economics predict that as hunted species become rarer and harder to find, hunters will switch to easier prey. The cost of investing in the hunt no longer covers the price on the market. Fishermen who have to go farther out to sea only to haul up puny catches will reach a point of weariness and withdraw, and so the fish will not go extinct; a low equilibrium will be established, beneath which populations can recuperate. Once luxury consumption enters the picture, however, the models fall apart. Now rarity per se is at a premium. The scarcer an animal, the higher the price it fetches on the market, the more attractive the commodity becomes for affluent consumers who wish to stand apart precisely by dint of its expensiveness, and as long as there are individuals ready to pay any price for the last specimen, hunters will go for broke. When consumption is fuelled by affluent segments with an insatiable preference for prestige, the presumed protection for species breaks down, and they are instead pulled into an 'extinction vortex', which is now essential to the workings of the global wildlife trade.

In the market of bourgeois theory, demand tapers off when prices rise. Here it works the other way around. Suppliers may even have an interest in frivolous wastage: traffickers of rare parrots from Indonesia were unbothered by smuggling methods that killed 90 to 95 per cent of their goods. They would stuff the parrots into plastic bottles, throw them into the sea and pick them up on open waters, and the circumstance that at least nine out of ten parrots died in the bottles *boosted* profits by making the species even rarer. The common pattern is that globally connected middlemen annex a line of hunting to circuits of capital accumulation. Traditional hunters are then squeezed out by, or occasionally promoted into, professional full-time poachers: the work transformed into wage-labour. The more destructive the vortexes, the more capitalist the hunting must become. When forests are emptied, sophisticated technologies are required to track down the last specimen, or to move on to fresher species and habitats.

We have here travelled some way from the bush of hunter-gatherers. But we have come closer to the place where the most recent chapter in this saga began: the market in Wuhan.

Made in China

If there was one feeling scientists working on zoonotic spillover did not express when Covid-19 took off, it was shock. A pandemic bursting out from bats is 'just a matter of time', one team concluded in 2018 – we shall encounter more such prophecies – but the exact location could not, of course, be predicted, just as no one knows where the next wildfire will hit. China has long been a top candidate. In what must be one of the more clairvoyant articles in the annals of scientific publishing, a large team of Chinese and international experts in November 2019 put out a study of the pool of bat-borne coronaviruses in which rural residents in southern China routinely dip their fingers – most often by touching intermediates: civets, badgers, chicken – and ascertained the potential for imminent spilling. The background to these concerns was, of course, SARS, that little portent of things to come, prefiguring the sequence of its successor by sending a bat virus into live animal markets where unsuspecting customers picked it up. When this is being written, one can only be as certain about the aetiology of SARS-CoV-2 as four months of peer-reviewed science permit. But the broad contours seem reasonably well established: once more from bats and into a market, although the identities of the stepping stones will perhaps never be known, as the stalls in Wuhan were thoroughly washed out as soon as the news broke. More to the point, however, is that *China could*

*become the cradle of this disease only because global tenden-
cies were present in concentrated form.*

Capitalist development has been as unkind to bats in
China as anywhere else: forest habitats going down for
eucalyptus plantations; cement manufacturers excavating
limestone hills and wrecking caves; tourists rampaging
through caverns; growing cities swallowing roosting sites.
Bats and other hosts have been under the typical pressure.
Meanwhile the human population has undergone massive
urbanisation – famously the largest migration in history
– and crammed into neighbourhoods ever hungrier for
meat. The consumption of meat rose by one third in the
first decade of the millennium. In this period of hyper-
accelerated integration in global capitalism, one niche
that expanded apace with everything else was 'wet markets',
so called because the stalls of animals are periodically
drenched with water after slaughter. In row after row,
these pulsating bazaars would display wild animals of
every kind, sourced from ever wider catchment areas and
catering to the curious taste.

The Chinese equivalent of bushmeat was no longer
what it used to be. Once more illustrating how adept capitalism is at co-opting and cashing in on pre-capitalist left-
overs – one more amalgam of archaic and contemporary
forms – traditional culinary and medicinal practices
involving wildlife turned into a for-profit industry, ruled
by the apex predator of universal equivalents in limitless
amount. The richest wanted the most precious commodities. Everything money can buy – 'Eating wild', writes

Quammen, 'was only one aspect of these new ostentations in upscale consumption, which might also involve patronizing a brothel where a thousand women stood on offer behind a glass wall'. Amid the Covid-19 pandemic, one group of Chinese ecologists headed by Jingjing Yuan of Xiamen University lamented the transformation of this selection of meat from protein supplement to token commodity of 'yuppies or tyrants because of the rarity and high price of the wild animals'. One study from the interlude between SARS and SARS-CoV-2 reported that consumers of such animals were predominantly young people with high incomes, not low, not middle. The typical customer was a young, rich, highly educated man. Patrons of wet markets testified to the joy of strolling among fish swimming in water tanks and birds squawking in cages – a 'lively and boisterous' shopping experience for all senses. The medium conjuring this world up in ballooning volumes was, as ever, money.

And the wet markets were seedbeds of zoonoses. Racoon dogs, flying squirrels, badgers, bamboo rats, crakes and assorted flavours of crow stacked on top of one another: no better way to start a riot of pathogens. The civets of SARS were captured in the Himalayas and presented as live commodities in Guangdong, but they might well have been infected *in situ*, for bats were sold in the same markets. A number of other animals tested positive too, having caught the virus in the stalls where faeces and urine and saliva inevitably mixed. SARS did not put a lid on this steaming cauldron. Ten years after it broke out, 'eating wildlife as food, purchasing ivory or big cats' pelt as crafts and souvenirs, and dressing in animal furs had become a fashionable lifestyle and symbol of elite status', the vortex spiralling on.

One prominent victim was the pangolin. This is the only mammal wholly covered in scales, seemingly slipping between taxonomic categories; with an armour reminiscent of a fish, it shelters in hollow trees, claws into termite mounds and curls into a hard ball when threatened. In the new millennium, the trade in pangolins revamped non-contemporaneous notions of its scales and meat as therapeutic, curing everything from acne and cancer to infertility, and put it on the menu as a fabulously exotic stir-fry. In the 1990s, China was still 'self-sufficient' in pangolins, but after 2000 the market demanded that populations be vacuumed up throughout Southeast Asia. By the early 2010s, the region had been practically emptied. Traders then turned to Africa and began plundering its stock, or recruited tribes in the eastern Himalayas that had long hunted pangolins on a traditional scale. In the 1990s, pangolins went for $14 per kilogram on the Chinese market. In 2016, the price averaged $600, reaching $1,000 in some restaurants – which only made this commodity even more irresistible to a substantial number among the *nouveau riche*. Propelled by such concoctions as pangolin wine, a supposedly health-conferring elixir produced by boiling rice wine with a baby pangolin, the species is now in free fall. Some early reports suggested, as we have seen, that it detonated SARS-CoV-2 on the

Wuhan market. One team claimed to have mapped a roughly 90 per cent identity between the RNA sequences of a virus in caged pangolins and those of SARS-CoV-2; others put it closer to 100 per cent. It remains to be seen if the genetic dissections can confirm this genealogy.

Other intermediate species have, as of this writing, been proposed: the bamboo rats sold on the Wuhan market came from an area inhabited by corona-carrying bats. In 2018, these rodent delicacies were taken from the wild and brought into farms owned by the Huanong Brothers, who became Chinese internet celebrities, model treasure hunters for others to emulate, when they posted films showing how to cage bamboo rats, feed them, torment them, slaughter them, eat them, sell them and get filthy rich. Bats may well have paid these farms a visit. Another tabled scenario is that uncommodified bats gathered at or near the Wuhan market of their own volition. There is a bridge in the city, the Yangtze River Bridge, decked with rows of green light that shine throughout the night and attract bats on the move; roosts have been observed around this structure, which is twenty minutes away from the market site. The bridge bats could have interacted with other animals in the neighbourhood, or themselves gone to the market to forage among the ample supplies of insects. All of these storylines rest on circumstantial evidence. But the clues so far point to the Wuhan market as an 'incubation bed' for coronaviruses that can mutate at the evolutionary speed of light; while it took eight million years for the genome of the human species to evolve by 1 per cent, RNA viruses of this kind can accomplish the same run in a matter of days. Stuff wild animals on top of one another and all pandemic hell will sooner or later break loose, as a predictable by-product of how the wild is handled.

Now there is nothing uniquely or exclusively *Chinese* about this handling as such. Consuming the wild as a luxury delicacy or trophy appears to be a transhistorical habit of dominant classes. Anyone moderately familiar with Egyptian pharaohs or English lords will recognise it. There are now more tigers held captive on private properties in the US than there are in the wild, in the world. Owners of ranches in Texas have a particular penchant for flaunting their wealth through one of those big rare cats. (This phenomenon entered mass culture when Netflix in late March 2020 threw the documentary series *Tiger King* to an audience hungry for entertainment in lockdown, producing some unintended irony and circularity. During a pandemic rooted in the caging of wild animals, people confined to their homes became glued to screens showing the 'murder, mayhem and madness' of people caging wild animals.) After China, the US is the second market for illegal wildlife commodities, but much of it isn't even illegal: pangolins are still traded in the open, the Trump administration refusing to even respond to the demand to put them under the Endangered Species Act. Whether the trade proceeds under the table or not, it is, as Vanda Felbab-Brown observes in *The Extinction Market: Wildlife Trafficking and How to Combat It*, rich people that keep it going.

The rule is very general. Rich Saudis pluck rare ante-
lopes from Somalia and snow leopards from Afghanistan;
rich Mexicans love boots made of reptile or crocodile skin;
rich Russians have developed a new fondness for furs. In
Europe, the market caters almost exclusively to luxury
shoppers. It also deals in food. In the last years BC, one
could find zebra steaks in Germany, crocodile sausages in
Norway, marsupials and camels and pythons in the meat
halls of Sweden – imports exploding in the 2010s, the
pythons going for 120 dollars per kilogram – not to
mention whale in Japan or turtle in the US; in California,
white abalone entered the vortex in the early millennium,
its population driven down by 99.99 per cent. The extinction market is part of how the one per cent lives, not the
essence of any national culture. What really set off the
vortexes in China was precisely the integration of the
People's Republic into globalised capitalism: circuits of
capital spinning in the markets, wildlife from all continents newly accessible through the ligaments of trade.

Nor is there anything uniquely and exclusively Chinese
about cheap labour. Nor about coal. But that doesn't mean
that these things cannot constitute serious problems in
China, only that they must not be treated in isolation from
the rest of the world, and when learning to see that world
in China and China in that world, there is no better place
to start than *Severance* by Ling Ma, published in 2018,
surely the most sibylline plague novel of this millennium.

Candace Chen has landed a job in New York City. The
daughter of Chinese immigrants, she is hired by a

publishing house to oversee the printing of books in Guangdong. She tours factories where workers stand with earplugs next to whirring and grinding machines and is subtly reminded that she and her company 'undercut the value of our labor year after year'. But something is brewing in the manufacturing zones. A mysterious lung disease spreads through the factories, forcing the suppliers of Candace's company to shut down. It turns out to be 'Shen Fever', an infection 'contracted by breathing in microscopic fungal spores', often mistaken for the common cold; early symptoms include headaches and shortness of breath. In its later stage, the fever is fatal.

Soon the numbers are on the uptick in the US too. Americans try to go on with their daily lives and keep the tally at arm's length, but the scale of the disaster eventually sinks in. In New York office buildings empty, employees switching to remote work from home. Designers put their logos on face masks. Gradually the city is vacated, bus services shutting down, supermarkets closing, Times Square standing deserted – 'There were no tourists, no street vendors, no patrol cars. There was no one.' (Some also leave after repeated hurricanes.) As she roams the streets and snaps photos of the empty city, Candace wanders into a half-open department store and randomly picks up a Victorian-style lavender teddy and glances 'at the label sewed in the back: Made in China. Of course it was.'

Having entered the US in imported goods, Shen Fever rips through the population. 'No matter where you go, you can't escape the realities of this world.' An exercise in cognitive mapping of the highest order, *Severance* does not, of course, correctly foresee all the details – it would have been pointless and impossible – but the novel deserves to be read as every bit as prescient as half a dozen papers on zoonotic spillover in *The Philosophical Transactions of the Royal Society.* Before it happened, *Severance* zeroed in on the umbilical cord between the next 'Chinese virus' and fully globalised capital.

If parasites had wings

The factor that enabled SARS-CoV-2 to spread from China to the rest of the world was global transportation. This is, indeed, the *sine qua non* of a pandemic – a spillover event alone is not sufficient; there must be a grid through which the pulse can propagate. If a hunter from an uncontacted tribe gets infected by something and dies in his sleep, that's the end of the story. It follows that the early modern history of infectious disease is written in the ink of merchant capital, specialising in buying cheap goods in faraway locations and traversing the globe to sell them dearly. In *Contagion: How Commerce Has Spread Disease*, Mark Harrison chronicles the epidemics recorded in the ledgers of trade, starting with the Black Death: a pest bacterium jumping from marmots to rats and then aboard the ships of Genoese merchants collecting Chinese silks and spices in Levantine ports. When those traders set foot in Sicily in October 1347, the plague arrived in Europe.

Of even greater consequence, of course, was the landing of Christopher Columbus in the Caribbean, which brought the panoply of diseases from the Old World – smallpox, measles, typhus, influenza, probably also pneumonic plague – to the Americas, whose inhabitants had never encountered those germs. Wave after wave of infection cleared out the hemisphere for the colonisers.

As cataclysmic as these events were, however, infectious disease still operated under a restraint: it could only travel as fast as the wind blew. Up until the early nineteenth century, it took one year to circumnavigate the globe. If a voyager carried a virus onto a long-distance sailing ship, he or she would either die or recover and cease to be infectious before arrival. But this restraint was shattered when ships were harnessed to a fossil fuel.

A steamship could outrun the wind and plough through the waves, since its fuel bore no relation to these forces: it came from under the ground, power stored from the distant past. Coal opened a gap in the barrier hitherto holding pathogens back. When steamboats were launched in international waters in the 1830s, the duration of some trips was halved, and then halved again in the coming decades, as the technology relentlessly improved. Now infected travellers could disembark without even having developed symptoms. The first steam-powered epidemic broke out in 1844, when *Eclair* returned to Portsmouth with an ailing crew; the boat had travelled up and down the coast of West Africa to further British trade and in the process acquired yellow fever, a disease that could barely survive the sluggish journey of a sailing ship. The arrival of *Eclair* activated the anxiety over the new prime mover still prevalent in Britain at this time: one expert on the fever believed that it had been imported through the 'artificial warm climate having been kept up during the voyage by the fires of the steamers'. The heat of coal combustion created a 'greenhouse' for infections, an engine room so warm as to sustain tropical infections all the way to northern latitudes where they did not belong. Merchant ships carrying coal as their commodity were thought to be particularly pestilent.

In the age of steam, infections were handed new schedules and itineraries: yellow fever took the steamboat over the Atlantic for the first time in 1849, setting off frightful epidemics in Rio de Janeiro and New Orleans. In 1832, a cholera pandemic engulfing Europe crossed the pond, followed by an outbreak in 1848 and another one in 1866, said to spread with 'unexampled rapidity' from Mecca to Mississippi. In the 1890s, a plague exited Guangdong through the arteries of the opium trade – itself originally secured with the aid of armed steamboats – and diffused throughout the world, calling at Sydney and Santos. Measles would never have made it to Fiji with sail. If the indentured labourers ferried over from India to toil at the sugar plantations there brought the disease on board, they would either have been fully recovered or committed to the deep when the three months' voyage came to an end. But once the British switched to steam and cut the crossing down to one month, the virus could step ashore in

Fiji. None of these later epidemics became reapers as grim as the Black Death or the Columbian invasion. Their historical distinctions pertained not to casualty tolls, but to dimensions of space and time: these events were global and fast, to a degree no disease had been prior to the rise of steam power. And they were harbingers of worse to come.

In 1918, a virus spilled from an avian reservoir host, probably wild ducks, to humans, probably in Kansas, and shot out across the world on board steamboats. It became known as 'the Spanish flu' because it was first reported in the Spanish press, free from the censorship imposed on the countries fighting out the last stages of World War I; more accurate would be the American or even the steam flu. In three great global gyrations, it killed at least 50 million people over 18 months, the most concentrated episode of mass death in human history. The second was the worst. In August 1918, the virus mutated into a strain that penetrated deep into the lungs and coloured the skin blue and made a quarter of its victims drown on frothy, bloody fluid welling up from inside. This strain struck at almost exactly the same time in three ports: Boston, Brest and Freetown. The third, capital of Sierra Leone, was a pivotal coaling station for oceangoing steamers. (The British had opened mines in the area.) The laconic log book of the *HMS Mantua* says that the ship had four men on the sick list when it left Plymouth on 1 August. The list had grown to 124 when it arrived at Sierra Leone after a fortnight – 'Commenced coaling; native labour.' Five days later, the log records the first death, a seaman passing away from pneumonia. The seamen started dying by the dozens, and as the 'natives' loaded *HMS Mantua* with coal, the influenza entered the African continent, first killing the colliers and stevedores of Freetown and then taking the railway inland; more disembarkations soon followed in other ports, the virus darting into the continent on trains. It was 'as though the colonial transportation network had been planned in preparation for the pandemic', two historians of the event note. Some two million died in sub-Saharan Africa, likewise the most intense, temporally concentrated demographic disaster in its history. Steam on sea and land took the pandemic to the four corners of the world – no place escaped it, except a few isolated islands with rigid quarantine measures in place – leading one historian to conclude that 'it was, in sum, a pandemic driven by the steam-engine'.

But steam navigation is outshined by aviation, the highest stage of fossil-fuelled transportation so far. High-octane gasoline can conduct infected humans to their destination in a fraction of the time coal could manage. Indeed, transcontinental flights are so fast that some vectors of pathogens – notably arthropods – can survive them undetected and steal away; more commonly, human globetrotters take a virus from one continent to another within hours, barely cutting into incubation time. More airports mean more points of entry. More flyers mean a larger susceptible swarm; more planes in the air a greater number of hothouses, where passengers are exposed to the germs from that coughing fellow over there and can do nothing about it, least of all let in some fresh air. This 'viral superhighway' in the troposphere has expanded as relentlessly as steam navigation once did, on a higher level. While total human population has grown sevenfold in the last two centuries, the mobility of its Western part has increased *a thousandfold*, half of the growth occurring since 1960 – like the second half of the nineteenth century, but on another order of magnitude.

SARS first demonstrated the performance of the superhighway. Over the course of hours and days, the virus from the bats and civets and markets flew into Toronto and Hong Kong and Singapore, the only good luck being, as we have seen, that symptoms preceded peak infectivity. SARS-CoV-2 shook off that last restraint too. Once it had spread through Wuhan, a transportation hub of 11 million people, it stole out on a plane to Bangkok, site of the first known overseas case, and farther on to Tokyo, Seattle, Seoul, Stockholm, where arriving passengers seeded independent outbreaks. By late January 2020, the virus had become an unstoppable sponger on the global aviation network.

But this network is not, of course, evenly distributed across the world. It is rather thin in Baluchistan or Somaliland. Flying is something that rich people do more than others, which explains the peculiar timeline of victimhood in this pandemic. Conversely, the spillover would never have reached that far if it had occurred in, say, the Democratic Republic of Congo rather than in the

People's Republic of China, whose place in globalised capitalism and *ipso facto* in aviation is special. No frightened bat and no connoisseur of pangolin wine could ever spread anything outside of the grid. If the Spanish flu was propelled by coal and steam, Covid-10 was powered by oil and aeroplanes, the common denominator plain to see.

Towards a theory of parasitic capital

With this, we are still far from exhausting the drivers of zoonoses and their potential pandemicity. One could mention ecotourism, which has its own valorisation of rarity, if with the best of intentions: visitors from afar wanting to come as close as possible to critically endangered primates. Can we touch them? Dams can become abodes for mosquitos. Excessive use of pesticides and antibiotics can cascade down the food chains to set pathogens free – in India and Pakistan, overprescription of diclofenac poisoned vultures, causing one of the fastest collapses of bird populations ever recorded and leaving uneaten carcasses as breeding grounds for parasites – and making microbes resistant. And then we still have not touched on the livestock industry. The practice, entirely novel in history, of forcing thousands of inbred animals under one roof is no art of healing. As Rob Wallace and others have detailed, animals pressed close together can easily become stressed and shed their microscopic guests; because they have been standardised into genetic monocultures, there are no buffers or 'firebreaks' against infections; viruses have free run of packed farms and can use the waste lying around or running off as a springboard. Outstretched commodity chains have the capacity to sustain transmissions across thousands of kilometres.

Here it is domesticated, hyper-exploited animals that form artificial lagoons of pathogens. But they can inherit viruses from the wild, such as from bats hovering over pigs (Nipah) or camels (MERS), or from migrating birds deprived of their old stopovers when wetlands disappear and they are driven to drop in at poultry farms. The latter is one scenario of the dreaded avian flu, described in hair-raising detail by Mike Davis. So far, the livestock industry has a track record of dozens of outbreaks, none coming anywhere near the virulence of Covid-19, in which it has not been implicated. But next time it could well be the farm animals' turn. With its copying of the US-style 'livestock revolution' – billions of pigs and cows and chickens confined in colossal facilities to feed the desire for meat – China is, again, a top candidate.

It thus comes across as a little clueless to suggest, as the above-quoted German liberal pundit did in March 2020, that 'a virus pandemic has just one cause. Climate change, on the other hand, is a highly complex issue.' If anything, the reverse seems truer. It would then seem equally foolish to deduce all these variegated drivers of pandemics in our time from a single source – and yet one cannot ignore the reappearing imprint of one meta-driver. Capital abhors the vacuum of wild nature. The capitalist class, we might recall, was brought up on hatred towards it. The pre-eminent philosopher of the plantation, John Locke, expressed the feeling eloquently: in his scheme of things, the original condition of the world was that of an unredeemed 'wild Common of Nature'. The mission of human beings, or more precisely property-owning human beings, was to abolish that condition. The wild common ought to be enclosed, rendered productive, improved – in short, converted into a fount of profit. 'Land that is left wholly to Nature, that hath no improvement of Pasturage, Tillage, or Planting, is called, as indeed it is, *Waste*; and we shall find the benefit of it amount to little more than nothing.' Wild nature is worthless waste – an abomination in the eyes of capitalists, because it is a space of resources that has not yet been subjugated to the law of value. What Locke put on paper has since become not so much a proud philosophy in the heads of the bourgeoisie as the *modus operandi* of accumulation as such.

When capital encounters wild nature, it does not step back to admire it or pay its respect. It does not take a walk and return home with fond memories, or with food to fill the belly. No such actions exist in its behavioural repertoire. Capital can relate to the wild *only by attaching itself to it*, so as to make it give up commodities that bear exchange-value; and in the very moment of its success, that nature is wild no more: it is clear-cut, captured, caged and carried off to the market. It is eaten from within, but the damage is inadvertent. Capital doesn't *mean to* destroy the intricate cellular structures of wild nature; it doesn't have an intention formed in the mind and then engage in efforts to realise it – there is just no other way for it to replicate. The fastening and sucking are in its DNA (or is it RNA?); the moment they cease, the reproduction of capital is over. Unlike other parasites, this one cannot stay content with vegetating in the furs or veins of other species for millions of years of co-evolutionary equilibrium. It can subsist solely by expanding and, in this sense, it exhibits a sort of permanent pandemicity; it doesn't return to lurk in the shadows until the next visitation, like Ebola or Nipah. Once it had leapt out of its reservoir host on the British Isles, it commenced the long historical work of subsuming wild nature on this planet, be it in the form of a palm oil plantation, a bauxite mine, a wet market or a rat farm. All of these and uncountable other entities represent wild nature dragged into the chain of value, and *given the biological fact that pathogenic microbes are constituent elements of such nature, capital must call them up too.* It cannot avoid splashing around in the pools of pathogens, like a gold-digger in the river mud. But such a hypothesis of capital as meta-virus and patron of parasites must, of course, be backed up by much further and deeper research.

On a slightly lower level of abstraction, we can propose the following theorem: time-space appropriation plus time-space compression equals high risk of zoonotic pandemics. Capital grows by dilating its material throughput. The more biophysical resources that can be processed into commodities and sold, the greater the profits; the greater the profits, the more resources can be acquired, and so on. Capital takes hold of land where the resources sprout – a law of a tendency with few countervailing forces that can be read off from aggregate data: in the year 1700, 95 per cent of the planet's ice-free land was either wild or modified and used so lightly as to be categorised as 'semi-natural'. By 2000, the proportions had been reversed. The better part of the land was now subject to domination of varying intensity, the largest wild pockets found in inhospitable desert and tundra, tropical forests perforated fast. For historical reasons, this arrow of appropriation has shot out from Europe – most of whose land had become dominated already before 1700 – to the rest of the world, much of it within the circles of Cancer and Capricorn. Land is hauled in, forming an immense collection of commodities. But it is not the only source thus impressed: labour too is embodied in trade flowing north. While the former is of greater immediate biological and epidemiological import, the latter must be included for the nature of the process to come into view.

In a sign of the times, research on unequal exchange nowadays focuses on ecological variables. But the papers reporting on vertical species loss dovetail with articles demonstrating how labour is similarly transferred through trade. One of the most remarkable calculates how much labour countries import to satisfy their demands: measured in full-time person-years of employment embodied in commodities, hundreds of millions of lives' worth of labour are shifted across the global marketplace. One resident in Hong Kong relies on seven workers – or 'servants' – from the rest of the world, in addition to the domestic workforce, to produce the goods he or she consumes; in absolute numbers of hours, the United States is, of course, the top importer. At the opposite end stands Madagascar. It needs less than one third of its own workforce to make what it consumes, while more than two thirds toil in producing things enjoyed elsewhere. As of 2010, the top seven exporters of embodied labour were Madagascar, Papua New Guinea, Tanzania, Tajikistan, Cambodia, Zambia and the Philippines: some direct overlap with the biodiversity drain. If capital needs biophysical resources to make profits, it relies on labour to process them at high rates of exploitation – the original intra-species parasitism – which forms another abiding reason to go south, where the two tend to blur into one phenomenon. Alf Hornborg has called it 'time-space appropriation'.

And the appropriation of time and space proceeds through their compression, as capital seeks to shorten turnover time: the faster an investment can be recouped, the quicker commodities are sold and return their yield, the greater the profits. Capital grows by higher velocity as well as by higher volumes. This, as David Harvey has theorised it, is 'time-space compression', the equally lawlike tendency of spatial barriers being broken by ever-faster technologies 'so that the world seems to collapse upon us' as impulses are transmitted across the globe with near-absolute instantaneity. When he introduced the concept in *The Condition of Postmodernity*, Harvey illustrated it with a map of a world shrinking in three steps. First there was the big, roomy earth of 1500–1840, when the fastest horse-drawn coaches and sailing ships managed ten miles per hour; then steam-powered locomotives and boats supervened at sixty-five and thirty-six miles per hour respectively, effecting the first round of compression; then came propeller aircraft, followed quickly by jet aircraft that shrank the globe to a microchip. Now if we tie together these two forms of capitalist production of time and space we end up with a recipe for infection. Capital is fastened to ever more land and sucking its contents into circulation at an ever madder pace, and this must, as a general law, result in a high risk of zoonotic pandemics, as one consequence of the ecological havoc caused. But only one among many.

How corona and climate differ: second cut

It should now be abundantly clear that the comparison between the climate crisis and Covid-19 rests on a category mistake. It's a bit like comparing a war with a bullet. Covid-19 is one manifestation of a secular trend running parallel to the climate crises, a global sickening to match the global heating. In March 2020, climate advocates were keen to point out that once this disease dissipates, the earth will still be warming up and send more thunder our way. That is true, of course, but once this disease has run its course, *the earth will also send more pestilence our way*. There is nothing to suggest that Covid-19 will be the last of its kind, as the rate of interaction between the human economy and 'pretty much all potential reservoir species' is rising steeply, driven up by some of the very same forces that impel the inexorable sixth mass extinction. One of the few scientific papers to methodically compare anthropogenic climate change and zoonotic spillover – we shall refer to it as 'the Pike paper' after its lead author – infers that in the coming decades, under a scenario of business-as-usual, the number of 'emerging infectious disease events' in the world will rise by *more than five per year*. Not all of them will, of course, reach as far on the dark side as Covid-19. But as experts from Quammen to Wallace pointed out when the latter started raging for real, this disease will, under said scenario, with a statistical probability approaching certainty, have successors. Some of them could be worse. One virologist announced that 'we are in an era now of chronic emergency'.

Ears have been as deaf to the science of spillover as to that of climate, if not more so. The alarm signals have echoed without being heard, as in a forgotten cave. Already in 1994, the eminent dialectical biologist Richard Levins and his colleagues warned that 'creating new habitats – for example, by bulldozing forests – permit rare or remote microorganisms to become abundant and gain access to people'. 'How much worse will things get in the tropics as biodiversity declines there?' asked one team despondently in 2006. 'As the line dividing human and wild habitats becomes thinner, we might be brewing the world's next big pandemic', ran a typical preamble in one of the *Nature* journals in 2017; more specifically, a group of chiropterologists studying the effects of deforestation deemed the risk of a coronavirus pandemic 'very high'. The apathy in the sphere of policy-making was bewailed in a recognisable manner. There will be disaster 'unless there is a major global paradigm shift towards sustainability' – or, 'Can we reverse or mitigate the trends before we're hit with another pandemic?' as Quammen asked in *Spillover*. In an early analysis of Covid-19, Wallace looked back on the virus strains coursing through the young millennium and recapped the answer given in practical terms: 'And near-nothing *real* was done about any of them. Authorities spent a sigh of relief upon each's reversal and immediately took the next roll of the epidemiological dice, risking snake eyes of maximum virulence and transmissibility.' These attempts to shake the putative top guardians of public health out of their slumber had no institutional trappings equivalent to those of climate science, no IPCC or UNFCCC. But the exasperation was all too similar.

We must then reconsider the character of the actions undertaken by states in the spring of 2020. Their closest analogue might be the reaction of the Australian state apparatus to the bushfires: declaring a state of emergency in the worst-hit areas and drafting the largest military units since World War II to cope with it. In early January, naval ships evacuated thousands of citizens huddling on beaches beneath the crimson sky, while infantry assisted firefighters trying to contain the blazes and one minister upped the war analogy: 'This is not a bushfire. It's an atomic bomb.' In both cases, actions were palliative. They treated the symptoms once they had blasted through the bodies of populations *and capitalist states seem incapable of anything more* – at least of their own initiative – save a modicum of preparatory adaptation, as in building sea walls or hospitals; here too, however, the shortcomings have so far vastly outweighed the achievements. When it comes to mitigation, as in not letting the pathology run its course, nothing is in sight. This does not obviate the spectacular character of some of the interventions made against Covid-19 or their possibly inspirational value, but it puts them in perspective.

What, then, remains of the differences between corona and climate? There are indisputable disparities in space and time. Global heating accrues through the atmosphere and flares up on earth as extreme weather events or other perturbations, which have, somewhat paradoxically, so far stayed on a local or regional scale: fires in Australia, floods in Iran, drought in central Chile (entering a grievous tenth year in 2020). Each delineates its own arc of misery. With Covid-19, global sickening, if that is an appropriate term, has surpassed the heating in its ability to jump scale from locality (a wet market in Wuhan) to globality in virtually no time. There are, of course, simple biological reasons for why a virus can be universalised so fast. A farmer whose land withers in Chile does not send off any organism to run riot through his species. Once the spillover has occurred, a pandemic is a human affair, broadcast through interpersonal contact at eye level, whereas impacts of climate breakdown will always be mediated through the material substrata of human existence. The analogue on the climate side would perhaps be simultaneous bad harvests in several dispersed breadbaskets, shortfalls so severe as to make the global food system buckle, or a major reorganisation of the thermohaline circulation or some other climate subsystem that ripples into every backyard. If continental ecosystems start to collapse, the domino effect will respect no borders. But so far, corona beats climate in earth-encompassing uniformity.

Global heating, on the other hand, is everywhere all the time; no forest or beach front can slip from its grasp, whereas infectious diseases come in bursts and fade. The background conditions that produce pandemics might be progressively intensified, but they are not manifest in the same unremitting fashion. Rather, the trend is one of increased risk or loaded dice, manifesting as discrete events that appear at intervals, between which there is nothing or very little. Even out of hurricane season, Caribbean beaches erode. One could also posit that global heating has an inherent potential for self-perpetuating deterioration – once we've reached four degrees we'll slide to six, which will ignite enough of the earth to take us to eight, and so on – whereas zoonotic spillover might follow more of an inverted U-curve. When the diversity of non-human animals nears complete annihilation, the pathogen load will go down. In the meantime, we are moving upwards on the curve. Biodiversity decline can certainly be self-reinforcing, with inbuilt tipping points and deadlines: the Pike paper estimates that business-as-usual closes 'the window of opportunity to deal with pandemics' in 2041. In the political present, then, the pending closures are similar.

And drivers are shared. The second major source of CO_2 emissions is deforestation. In the first decade of the millennium, it accounted for one tenth of the gas released into the atmosphere, dwarfed by fossil fuel combustion but outdoing it in one regard: while some 25 per cent of emissions from fossil fuels were embodied in trade – as in Sweden importing goods from China, where coal was burnt for their manufacture – the share stood at 40 per cent for deforestation. In other words, far more of the gas discharged in the process of bulldozing forests was an epiphenomenon of northern consumption, as in beef, soybean, palm oil. Aviation remains a minor source of CO_2, in a state of growth – up to Covid-19 – exceeding all others. Between 2013 and 2018, total annual emissions from fossil fuels rose by less than 1 per cent, while those from aviation grew by nearly 6 per cent, a rate tantamount to the building of fifty new coal-fired power plants every year. The US and China led the airborne pack. Pathogens thus have more than a tangential relation to certain drivers of climate breakdown, which, in turn, offers fresh opportunities to the former.

It has been well known for some time that global heating marks a long season of migration to the north and up to higher altitudes for wild animals. When the climate envelopes to which they are adapted begin to move, many have little choice but to try to follow. We now also know that this process is unlikely to be gradual: when a tolerance limit is breached at a site, it won't be one species feeling the unbearable heat, but whole assemblages of species that have made it their home for ages, suddenly undone. Such limit exposure is already underway in the Caribbean and the Coral Triangle and expected, under a business-as-usual scenario, to reach the rainforests of Latin America, Central Africa and Southeast Asia by 2050 or so. A lot of jars are waiting to crack in the sun and spill out their contents, which may, of course, be unable to adapt and perish, but not before attempting to track the liveable climate polewards and upwards. Such processions are already afoot, before the ultimate thresholds have been reached; here too, it will be a long slide punctuated by ruptures.

Along the way, the animals meet strangers. As they scramble for refugia, species without previous contact will cross paths and stage 'first encounters', perhaps share a range for a while, before having to move on and repeat the experience. For pathogens long inured to a relatively monogamous lifestyle, these will be moments of promiscuous licence: flocks of new animals filing past, like a meat market restocked every morning – so many opportunities to jump. Most of the tens of thousands of novel pathogen exchanges anticipated along these routes will take place between one species of wild animal and another, but it will be a moving laboratory of genetic recombination, in which parasites may learn to make longer jumps: and their hosts will bump into, or skirt past, humans. Viral sharing events are likely to be most common in places with fairly dense human populations, such as the Ethiopian highlands, Indonesia and – crossroads again – eastern China.

They might precede the most extremely disruptive temperatures. Nascent research on how global heating amplifies zoonotic spillover indicates a stronger effect during the first few degrees Celsius; five or six might cook too many hosts to extinction.

Among the animals in transit will be arthropods. Some act as vectors for infectious diseases, such as malaria and dengue fever, yellow fever and Zika – but they will not necessarily travel under duress, for as a general rule, a hotter climate is beneficial to them. It allows them to expand their ranges. There is no dengue fever in Denmark: it's too cold for the vector mosquito. Hence a long-standing concern for climate science has been the proliferation of this class of diseases, but recent fine-grained research gives no unambiguous answer to the million-death question: whether total malaria transmission will increase or decrease over the coming decades of warming.

Similar uncertainties apply to locusts. They too like it hot, and aridification outside of their traditional desert bases may open up new prospects. Swarms avoid flying during the cold night. Higher temperatures permit them to take off earlier in the morning and fly later into the evening, covering greater distances; as the warm air rises, they might be able to glide with the winds over mountain ranges – the Atlas, the Alborz, the Zagros – no longer effective as obstacles. But all depends on precipitation. For locusts to benefit from the heat, there must be good rains, as in 2018 and 2019. Projections for the regions where they breed point to current trends ratcheting up:

further decreases in total precipitation, but with greater variability; long dry spells and then unexpected storms. All in all, 'we cannot exclude a higher potential risk of local swarming', particularly for southern Africa. (As these words are written, it has just been announced that a second generation of swarms has formed in the east of the continent, estimated by the UN to be twenty times larger than the first of 2020. 'This [swarm system] is very active, destructive and we are worried it has come at the time of lockdown. We are a bit overwhelmed,' sighed a minister in Uganda.) But the animals of most immediate concern for the moment – or so it seems – are bats.

Among mammals, bats have a special facility for migration: they can fly. And climate change can mess with most aspects of their lives. When to ovulate and give birth, when to hibernate and for how long, what sites are suitable for roosting, when and where insects and fruits show up, where to find water – all is up in the air in rapidly warming niches. At some point the bats will fly away to cast about for safer grounds. While most of the map for the Chiroptera is still blank, a raft of studies has reported that populations and species are indeed on the move. Bats in Costa Rica have been relocating from the lowlands to the cooler cloud forests in the mountains; the Brazilian free-tailed bat is making its way through the Deep South of the United States, reaching Virginia in 2018; bats in the Brazilian Cerrado are being evicted by the combined forces of deforestation and heating; one species common to the Mediterranean has more than doubled its range in three decades, scurrying into northern France and Poland. And China goes the same way. One study sampled seventeen of the People's Republic's 130 bat species, including half a dozen kinds of horseshoe bats, infamous as the most competent corona hosts. It found a well-nigh universal trend: range limits shifting towards the north in the past half-century, on the heels of rising temperatures. This south-north gradient of migration could not be explained by land-use change, which does not conform to it – forests are not systematically logged from south to north – indicating that the heating might be *a more powerful disseminator of bats in China*. And then there are the modelling studies. Under business-as-usual – always the baseline for the worst-case scenarios – the double whammy of heating and deforestation will force up to 99 per cent of bat species in Southeast Asia to migrate by 2050. One per cent can stay in place. Many will not, of course, be able to cross into asylums, as they will smash into the walls of infrastructure – the final interface, as it were.

It doesn't require any advanced mathematical intelligence to infer the meaning of this for the future of spillover. It's rather as if the human economy had resolved to lift up the container of coronaviruses and other pathogens and pour the load over itself. Climate change is the supreme stressor: for bats, from one day to the next, insects absent when most needed, hurricanes flattening roosts, droughts forcing lactating females to fly longer for water – the kind of stress that has been observed to induce mass shedding. Some of the bats drifting to the Malaysian pig farms and dumping Nipah had fled the wildfires of 1997–8, kindled by drought from the worst El Niño of that century. An exceptionally prolonged dry season preceded the Ebola outbreak of 2014 and may have contributed to more frequent encounters between bats and people. Were some of the bats hanging out in Wuhan or on the paths leading there runaways from the heat? At this stage, we don't know. But it has been noticed that all three coronavirus epidemics so far in the millennium have been associated with dryness: SARS followed an epic drought in Guangdong; MERS was first detected in rain-free Jedda; SARS-CoV-2 erupted amid the worst drought in the Wuhan area in forty years. The hypothesis here is that *the coronaviruses themselves* prosper in low humidity. Like so many others, it has yet to be confirmed, but it should now be evident enough that corona and climate do not form separate, parallel lines. Corona can be an effect of climate; not the other way around. More importantly, the two are interlaced aspects, on different scales of time and space, of what is now one chronic emergency.

Wounded on the battlefield

In the last days of March 2020, news broke that a seventeen-year-old in Los Angeles County, the son of immigrants, had asked for urgent care at a hospital and been turned away because he did not have health insurance. A few hours later, Covid-19 killed him. More details subsequently emerged;

the boy's family did in fact have insurance, but misunder-
standings and bureaucratic mishaps appear to have slowed
down treatment. The story brought home a division by now
well apparent: there was one Covid-19 for the poor and
another for the rich. Although the unusual timeline quickly
involved the latter, the disease spread in the human population like water from a burst dam, released into a maze of
channels, some very deep, others narrow and shallow; in all
respects, inequalities determined how it unfolded.

Most conspicuously, some had access to excellent
health care. Others could only dream of sanitisers and
soap, let alone insurance and hospitals equipped with
brand-new ventilators. There were 12.5 intensive care unit
beds per 100,000 residents in badly hit Italy. The figure
for Bangladesh was 0.7; for Uganda, 0.1. Prior health
issues that weakened defences – diabetes, heart conditions, lung disease – divided along class lines; as for lungs,
they had already been worn down by the soot inhaled by
coal workers and neighbours of oil refineries (or wildfires).
Even the measures taken to turn the tide reinforced these
patterns. The rich could well afford some self-isolation in
private cocoons, a rather natural extension of their normal
lives. One form of aviation boomed: by mid-March,
bookings with private jet operators in the US had increased
tenfold, as clients took their families and private doctors
to vacation homes secluded far from contagious masses.
Real estate agencies in the UK reported a torrent of inquir-
ies for 'corona mansions', mostly from wealthy Londoners,
who were happy to pay £10,000 per week or more. 'These

are people who say to us "I'm going to be working from home, my wife is going to be working from home, it must be somewhere with a lot of space, large gardens and facilities such as tennis courts where we can relax."' Working people had that option rather more rarely. Many were compelled to keep going to work, including work in the health sector – meeting, testing, treating the infected. Self-isolation in a slum is a contradiction in terms. One category of places soon emerged as most defenceless of all: refugee camps, holding Rohingya in Bangladesh or Palestinians in Lebanon or Syrians in Greece in the most abject conditions (from original source to potentially deepest sink, the virus triumphed in spaces of mass confinement). In short, corona illuminated a fractal landscape of vulnerability, from the most global to the most local – and in this regard too, it was similar to climate. McKibben's law seems nearly as applicable here: maybe not the first, but most victims 'are those who have done the least to cause the crisis'. Indeed, if the rich were first to be hit – as in Brazil, where the super-rich flew the virus in and then shed it onto the working masses – the poor had no shields to hide behind. To understand why this seems to be the case along every front line of this chronic emergency, we might turn to theories of vulnerability.

At the most formal, anaemic level, vulnerability can be defined as the degree to which a system is susceptible to harm when affected by an external stressor. More graphically, the root of the term comes from the Latin *vulnerabilis*, used by the Romans to describe the state of a soldier lying wounded on the battlefield and therefore liable to receive the next blow as mortal. On this view, vulnerability is an antecedent of impact, a debility or injury inflicted by some humans on others before tragedy strikes. But it hasn't always been regarded that way. In the post-war decades, research on natural hazards – here we bracket diseases, for a moment – assumed that disaster was the result of an extreme geophysical occurrence: a flood, a storm, an earthquake. Vulnerability was a function of likewise geophysical features: the inclination of a hillside, the flatness of a coastline, proximity to a major fault. It was in the very nature of the events to produce harm. Causality ran in a straight line from the environment, with humans at the receiving end. This school – call it geophysicalism – ruled unchallenged in Western academia, until it came under fire in the 1970s from scholars working under the influence of Marxism.

The first salvo was launched in 1977 in the inaugural issue of a journal simply called *Disasters*. There Ben Wisner and his colleagues pointed to an increase in the number of victims of natural disasters, particularly in the Third World, and asked what might be behind this disturbing trend. Finding no rise in the number or magnitude of the hazards themselves – the droughts, the hurricanes, the volcanic eruptions struck to the same extent as before – they proceeded to argue:

> Then how does one explain the increased frequency and severity of 'natural' disasters? In short: by denying that these disasters are 'natural' at all. More and

> more people are becoming vulnerable to the occur-
> rence of certain physical events which have been
> occurring with a certain mathematically recon-
> structable probability for centuries, if not millennia.
> It is in the phenomenon of vulnerability – that is, on
> the human side of the man-nature relationship – that
> an explanation is to be sought.

Society, not nature: this is the death sentence for multitudes. Vulnerability, Wisner et al. contended, is really a
function of unequal ownership of resources. Instead of
viewing disasters as chance events or 'acts of God' that
irrupt into ordinary life, they should be seen as the starkest
truth about that life, whose inner structure they bring to
light. This is the cardinal idea of critical vulnerability
theory, elaborated in countless case studies: during a
drought in northern Nigeria, to take one classic example,
rich households stood the test thanks to the large size of
their cattle herds and other assets, whereas the poorer ones
bit the dust, meaning that the drought itself was at most 'a
catalyst' of selective pressure inhering in the property relations. Some owned the means for survival, others did not.

This school of thinking has received its most authori-
tative exposition in *At Risk: Natural Hazards, People's
Vulnerability and Disasters*, again written by Wisner and
colleagues, fleshing out the argument that what goes on in
nature is peripheral to the outcomes. Adversity is a fact of
life that comes and goes. Whether one can cope with it is
a matter of having enough land to farm, or adequate access

to water, or a stash of jewellery or a shed of tools to use in need, and this is exclusively 'determined by social factors'. Most fundamental of these are 'relations of production and flows of surpluses', as they decide what cushions an agent can dispose over. A population is divided into classes, and further into genders, ethnicities, age groups, citizens and migrants with antithetical positions: some wounded on the battlefield, others decked out in shining armour and ready for anything, including infectious diseases.

One chapter in *At Risk* is devoted to this type of hazards. Zoonotic spillover – a term not employed, although the phenomenon is discussed, SARS being a case in point – will become an epidemic or worse for people with inadequate diet, shelter, sanitation, water and access to health care, which might have been privatised or cut out of their reach. Wisner et al. are not, of course, the first to make this point; it tallies with a tradition of critical epidemiology likewise influenced by Marxism. In the same year as the original *Disaster* article, Meredeth Turshen attacked the paradigm of clinical medicine as excessively preoccupied with how the individual body reacts to disease, missing the bigger picture of class and other collectivities. She cited Engels's descriptions of how polluted air, poorly ventilated houses, overcrowded slums and omnipresent sewage predisposed the workers of Manchester to become ill. She could have also quoted Rosa Luxemburg: 'The doctors can trace the fatal infection in the intestines of the poisoned victims as long as they look through their microscopes; but the real germ which caused the death of the people in the asylum is called – capitalist society, in its purest culture.' Since the 1970s, critical epidemiology has agreed with critical vulnerability theory on emphasising the social over the natural: disease and disaster as produced through processes *internal to society.*

Seeping out from Marxism and into mainstream academia, the basic insight about vulnerability is now easy to find. Wisner et al. have made it enormously influential through what they call 'the pressure and release model'. Disaster strikes as a clash between two magnitudes: socially determined vulnerability from the one side, natural hazards from the other. In between, people are squeezed or 'crunched' as in a nutcracker. What truly accounts for the result, however, are the goings-on to the left, as depicted in the figure below. There is a 'progression of vulnerability', a sequence of causation running from 'root causes' via 'dynamic pressures' to 'unsafe conditions'. A capitalism plagued by deep inequalities (root cause) leads to corporate appropriation of land and accelerated urbanisation (dynamic pressures) that impoverish people and drive them to build homes on steep hillsides (unsafe conditions) – and then comes the deluge. The hazard is but a trigger that 'releases' the social pressures long accumulated: geophysicalism turned on its head. The political gist is plain. One solution, revolution; the root causes, at the extreme left of the model, must be dealt with through a 'revolution or major realignment in the balance of class forces'. Disaster planning 'must be, broadly speaking,

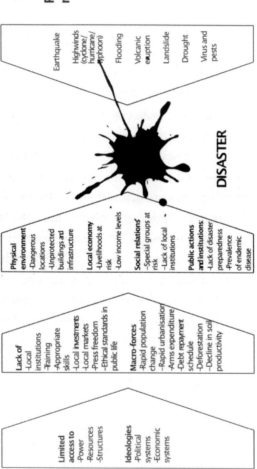

The progression of vulnerability

Pressure and release model

ROOT CAUSES	DYNAMIC PRESSURES	UNSAFE CONDITIONS	HAZARDS

ROOT CAUSES

Limited access to
-Power
-Resources
-Structures

Ideologies
-Political systems
-Economic systems

DYNAMIC PRESSURES

Lack of
-Local institutions
-Training
-Appropriate skills
-Local investments
-Local markets
-Press freedom
-Ethical standards in public life

Macro-forces
-Rapid population change
-Rapid urbanisation
-Arms expenditure,
-Debt repayment schedule
-Deforestation
-Decline in soil productivity

UNSAFE CONDITIONS

Physical environment
-Dangerous locations
-Unprotected buildings and infrastructure

Local economy
-Livelihoods at risk
-Low income levels

Social relations
-Special groups at risk
-Lack of local institutions

Public actions and institutions:
-Lack of disaster preparedness
-Prevalence of endemic disease

HAZARDS

Earthquake

Highwinds (cyclone/ hurricane/ typhoon)

Flooding

Volcanic eruption

Landslide

Drought

Virus and pests

DISASTER

After Ben Wisner, Piers Blaikie, Terry Cannon and Ian Davis, At Risk: Natural Hazards, People's Vulnerability and Disasters, second edition (London: Routledge, 2005), p. 51. Note that terms are reproduced verbatim.

socialist', so as to heal the wounds before people go to meet the forces of nature.

How, then, does this model incorporate something like climate change? In its early days, critical vulnerability theory had an uneasy relation to the scientific proposition of such change. In 1983, Wisner and colleagues derided 'much mystifying argument about climate change, especially following the prolonged drought over the African and Asian continents', for which they can be forgiven. The science was not as unequivocal in 1983 as it is now. But the crux of the matter is logical. The notion of climate change threatened the theory, because it removed the premise of hazards 'occurring with a certain mathematically reconstructable probability for centuries, if not millennia'. It sent turbulence into the right side of the model – the side those Marxists had wanted to get away from. If the hazards themselves redoubled in force and then redoubled again, it would be impossible to keep the 'root causes' of disaster to the furthest left. For the model to work, ironically, the storms and the floods, the landslides and the droughts *and* the viruses and the pests must be bracketed as chance events. Only then could focus be shifted to the perceived social side of the equation.

The theory, in other words, was not made for a warming world. In 2006, Wisner and colleagues attempted some ad hoc integration of this development, only to relapse into naturalising and downplaying it: 'Most disasters, or more correctly, hazards that lead to disasters, cannot be prevented. But their *effects can be mitigated* – what is otherwise known as adaptation. In *At Risk*, however, the view of climate impacts as themselves acts of God only treatable *post festum* is exchanged for a fuller acknowledgment of the new reality, whipping up entirely unprecedented amounts of drought and flood and disease. The revision culminates in the following sentence: 'With climate change, human action is responsible for both the generation of people's vulnerability *and* the increased level of hazard.' In the same moment, the model falls apart. The social is no longer on the left side solely. *It has saturated the hazards themselves.* It turns out that the model rested on an overestimation of the natural in nature, tucking away the actual hazards in a black box and setting to work on the purely social stuff – how people relate to one another. The error is as consequential for zoonotic spillover. As critical vulnerability theory once negated geophysicalism, a negation of the negation is now called for, with direct bearings on how to beat hazards like pandemics.

The dialectics of disaster

Since models of this sort are meant to be heuristic devices, we can allow ourselves some stylised simplifications. We want, first, a dialectical model of climatic disaster. Extending the artwork from Wisner et al., we might picture it something like this:

A dialectical model of climatic disaster

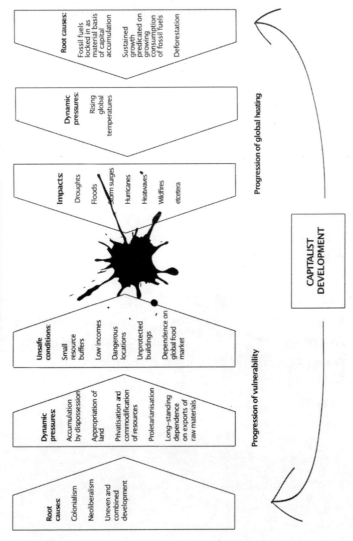

Root causes:

Colonialism

Neoliberalism

Uneven and combined development

Dynamic pressures:

Accumulation by dispossession

Appropriation of land

Privatisation and commodification of resources

Proletarianisation

Long-standing dependence on exports of raw materials

Unsafe conditions:

Small resource buffers

Low incomes

Dangerous locations

Unprotected buildings

Dependence on global food market

Impacts:

Droughts

Floods

Storm surges

Hurricanes

Heatwaves

Wildfires

etcetera

Dynamic pressures:

Rising global temperatures

Root causes:

Fossil fuels locked in as material basis of capital accumulation

Sustained growth predicated on growing consumption of fossil fuels

Deforestation

Progression of vulnerability

Progression of global heating

CAPITALIST DEVELOPMENT

We also want something similar for pandemic disaster. Without any perfect symmetry, it might look like this:

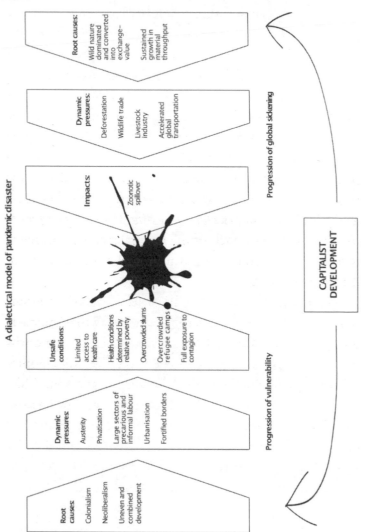

A dialectical model of pandemic disaster

Any number of factors could be inserted into these models, which are nothing if not sketchy. To further approximate the complexity of the realities, they would have to be entangled through recursive loops; zoonotic spillover, not the least, should be an impact in the model of climatic disaster. But the basic point that marks them off from both geophysicalism and pressure-and-release is that *similar social drivers are active on both sides of the equation* – not as omnipotent forces to which nothing can be added, but as non-negligible prime movers. If we imagine the models laid on top of each other, we can have a mental picture of the chronic emergency as represented in these two aspects – there are more, of course – and the emergency is *chronic and escalating by dint of what's going on to the right side.* The epoch of the Capitalocene is characterised by uncontrolled speed-up in the production of hazardous nature.

Where, then, has the political left focused its attention, and where ought it be? It is safe to say that most left discourse about Covid-19, at least a couple of months into the outbreak, stayed squarely on the vulnerability side of the equation. A typical sentence would sound like this: 'It is clear that people in Italy are dying not because Covid-19 is so lethal, but because the neoliberalization of health care and the EU's austerity measures are literally killing them.' A half-truth, it corresponded to a series of demands: immediate end to austerity, universal basic income to allow people to stay at home, universal health insurances in countries primitive enough to lack them, expropriation of tax havens to finance expansion of public health care sectors, higher payments for workers in hospitals and nursing homes, opened borders, vaccines developed with intellectual property seized from pharmaceutical companies . . . All of this and more are indispensable. Sadly, however, even a complete realisation of these demands would be insufficient *as long as the right side is left unaddressed*, and in much of this discourse, the virus itself was indeed seen as an act of God. It was a misfortune that befell humanity, and the rest is the history of class struggle. There were exceptions, of course – notably Rob Wallace, long a lone voice in the cave, who had another emphasis and suggestion: 'Let's stop the outbreaks we can't handle from emerging in the first place.'

One might regard Covid-19 as the first boomerang from the sixth mass extinction to hit humanity in the forehead. Once it did hit, however, issues of ecology still had trouble making it to the top of the agenda, which was mostly concerned with the pain of the concussion – again with exceptions: CNN reported that the well-established drivers of spillover 'suggest the need for a complete rethink of how we treat the planet'. Some people under lockdown in Bordeaux hung out a banner with the text *On est tous des pangolins*, 'We are all pangolins.' But overall, reactions were similar to those in a wildfire – where the hell are the firefighters? Why isn't the government doing its job? Ecological disasters have a way of overwhelming people's lives, so that all that comes to matter is survival, a moment not propitious for deeper contemplation or spontaneous mass overhauling of the material structures of society; and before the disaster hits, and once it has passed, business-as-usual just seems so very normal. When will it ever be put in the crosshairs? The most recent string of climate disasters, from the heatwave in the summer of 2018 (a nearly global event) to the Australian wildfires, indicates that there might be some prospects for elevated conscious-ness and action in moments of impact. Any tentative steps in that direction resulted from decades of hard work in climate science and movement. Coronageddon had no similar lightning conductors to hit, but given the power of the strike, rapid catch-up might, perhaps, be possible.

That would require a *What Is to Be Done?* for the right side. A left staying in its habitually defined social corner will only be capable of raising demands similar to 'sea walls for all' – better palliative action, but palliative. It will be overtopped. When the nature of the battlefield shifts this epochally, there will only be time to wash the wounds before new ones are slashed open. Any chance of getting *out* of the chronic emergency presupposes a different concentration of forces. To be 'radical', after all, means aiming at the roots of troubles; to be radical in the chronic emergency is to aim at the ecological roots of perpetual disasters. Corona and climate are not, it bears repeating, the sole components of this ordeal. There is a long list of time bombs waiting to blow up – insect collapse, plastic pollution, soil depletion, ocean acidification, renewed ozone depletion, not to exclude the possibility of nuclear meltdowns or other surprises – but the present has picked these two for a shortlist, and they are enough to keep us busy for some time. There is a point where they intersect with special intensity.

Fossil fuel extraction in tropical forests combines the drivers of climate change and zoonotic spillover in one bulldozer. Deep in the Amazon, the Brazilian oil company Petrobras pumps up oil and gas and feeds them into pipelines built by the Swedish company Skanska, now operated by the French multinational Engie; more pipelines are scheduled to be constructed. On the other side of the border, the Peruvian state is banking on an oil boom in the Amazon. So is the Ecuadorian, which has opened the most exuberantly biodiverse forest zone, the wet and cloudy Yasuni, with more species of insects, birds, mammals and amphibians per hectare than anywhere else in the world, expected to serve as refugia for animals fleeing drought and fires, to drilling. More than half of all oil exported from Ecuador goes to one single market: California. The extraction is financed by players like JP Morgan and Goldman Sachs. Fossil capital: parasitic capital.

On the other side of the tropics, in Indonesian Sumatra, the Harapan forest, inhabited by tigers and elephants and less charismatic species, is beating eaten away at the fringes by palm oil plantations, but the biggest threat as of this writing is a coal company preparing to cut a road straight through it to truck its produce. None of this, however, compares to the explosive charge of the plans to invade the peatlands of the Congo basin and extract some hundreds of millions of barrels of oil. This swampy rainforest area has long been known to constitute a 'virosphere' of rare luxuriance; in 2017, scientists could also demonstrate that it is one of the most carbon-dense ecosystems on earth, storing astronomic amounts of that element in the ground. One of the richest men in Africa, the well-connected Claude Wilfrid 'Willy' Etoka, wants to go in and drill.

This is all on hold during Covid-19. This will all come roaring back if investors have their way. Fortunately, they face resistance from indigenous populations and other local actors, currently best organised in Ecuador; but all projects of this kind, insofar as they come near realisation, grow out of the deep pockets of the North. They assign an immediate task for activists in that part of the world. For purposes of deterrence, the climate and environmental movements should here bring out the most militant tactics in their arsenal.

3

War Communism

There is a legitimate joke about Marxists having correctly predicted twelve out of the last three capitalist crises. But if any Marxist had ever suggested that the worst crisis in late capitalism would be precipitated by a virus from a bat, even the comrades most inclined to catastrophism would have shaken their heads and wondered what drug had induced this hallucination. (When these words are written, in late April 2020, the Bank of England has just flagged a 'contraction that is faster and deeper than anything we have seen in the past century, or *possibly several centuries*' – possibly, that is, in all the annals of capitalism.) There is, of course, the alternative of arguing that the mega-crisis now unravelling the world economy has merely activated the contradictions of overproduction, overaccumulation, overfinancialisation or the good old falling rate of profit – rather like the hazard in the pressure and release model – but this would be to place the undeniable proximate cause outside the system and mimic the bourgeois disavowal of it as an exogenous shock. But there is at least one Marxist crisis theorist with something to say to this moment. It is James O'Connor, who believed that capitalism does not have one Achilles' heel, but two.

When capital runs in the spiral of accumulation, profits and ever more profits dangling before it, it tends to invest in larger capacity than the market can absorb. It builds and produces too much, too fast for demand to keep up. If the rate of exploitation is high – if a very big share of the value produced by labour is wrung from it: real wages depressed, ordinary consumers cash-strapped – the problem will become acute. It might be papered over by a swelling credit system, keeping consumption going on borrowed means, but this will only build up financial instabilities and postpone the reckoning. This, roughly, is the 'first contradiction' of capitalism, for O'Connor: a tendency of capital to get ahead of itself and everyone else and therefore regularly fall down in spasms. It's a contradiction entirely internal to capital, something it cannot help doing, like a neurotic loner, conceived here in abstraction from its environs. This is standard fare of classical Marxist crisis theory.

But then, O'Connor continues, there is also a set of *'conditions of production'* of which capital must avail itself. Labour-power is one: there must be workers to do the work, and they must be reasonably healthy and fit for their tasks. Non-human nature is another. When capital attaches itself to both, as it must to get off the ground, it cannot avoid debilitating them, and the effects will become more injurious when it aggressively cuts costs and chases still higher profits, to the point where segments of labour and nature may disintegrate. Nancy Fraser, who has developed similar ideas into a much broader theory, speaks of capital 'free riding' on these 'background conditions' and spoiling them at the same time. (The general logic of malign parasitism.) Now, if this goes on unchecked, O'Connor argues, it will eventually recoil upon capital in the form of falling profits. A capitalist crisis *sensu stricto* breaks out. The property relations and productive forces tend to 'self-destruct by impairing or destroying rather than reproducing their own conditions'. This is O'Connor's famous 'second contradiction' – something that capital cannot, again, prevent itself from developing, but here the logic is more like that of the abusive man, who self-destructs by compulsively destroying those who give him life.

We can recognise this plot line in the present crisis. A sympathetic critic of O'Connor once objected, after the serene 1990s, that 'it is not true that the current era of capitalist development has generated a health-related contradiction around the availability of labour power'. Well, it has now. By pushing all the buttons of zoonotic spillover to the fatal eventuality – more generally, by altering the 'internal grammar' of nature, with Fraser – capital has deprived itself of a great deal of workers, for whom the workplace has come to seem like a pesthouse, and, no less fatally, of consumers, who avoid marketplaces like the plague, with the consequence that the wheels of accumulation come to a halt. This might be the first true O'Connor crisis.

A problem of the theory is the fuzzy mechanism for translating ecological woes to capitalist ones; O'Connor just assumed that it would happen, and so predicted – the thirteenth of the last three – that global warming would suppress profits and induce slumps. This has emphatically not happened *yet*, partly for reasons outlined above. The mechanism that constituted Covid-19 as a capitalist crisis – the one that has been absent for climate – was the *intervention of the capitalist state in a moment of relative autonomy.* States ordered the lockdowns. States enforced abstention from 'non-essential' production and consumption. It would thus seem that the second contradiction is actuated by the state protecting a background condition in jeopardy, in this case the bodily integrity of the population of producers and consumers. O'Connor himself hinted at such a scenario; now we have hard data. What he did not divine was the extent to which the second contradiction *would in turn actuate the first* – he kept them as separate; moreover, he expected that they would have opposite signatures. The first would make itself felt through collapsing demand. The second would instead raise the prices of things from nature and squeeze profits from the cost side (or, the first would hinder the realisation of surplus-value, the second its production). But in the event, the bullet with which capitalism shot itself in the second foot continued straight through into the first: and the system has never been more prostrate. Is this how it dies?

All roads lead to the seed bank

Marxists have proclaimed the imminent end of capitalism for as long as they have been around. Some have learned the lesson. 'The experience of our generation: that capitalism will not die a natural death', Walter Benjamin jotted down in *The Arcades Project*. The ones who truly suffer from a capitalist crisis are not the capitalists and their system, but the millions of working people – this time possibly billions – whose lives are wrecked. Indeed, every time another opportunity to finish off a temporarily weakened capitalism has been wasted, this mode of production has emerged stronger from the crisis (stronger in 1992 than in 1973, in 1948 than in 1929, in 1896 than in 1873 . . .). Outliving every host, capital is the parasite that never dies. Or is it? There might conceivably be some natural limits to its lifespan. Value cannot valorise itself on Venus. Sooner or later, even in the absence of state intervention to protect the background condition of a liveable climate, through some mechanism or other, axiomatically, a hothouse earth must begin to cause profits to evaporate. A planet of ruined conditions should be unable to yield monetary harvests, for the same reason that money does not grow on the other side of this tellurian atmosphere. One can imagine a chronic emergency that plays out in a concatenation of disasters: one pandemic after another, one climate impact after another, blow succeeding mighty blow until the foundations are too damaged and the whole system starts to totter. Something along these lines might have already happened at least once in history.

In *The Fate of Rome: Climate, Disease, and the End of an Empire*, published in 2017, Kyle Harper gives a new twist to the old story of the decline and fall of Rome. His tale begins with the rulers of the imperial capital staging one of their pageants of wild animals: boars and bears, elephants and elk, leopards and lions, all 'the glory of the woods' and the 'marvels of the south' paraded and then massacred in the arena just for the sake of it – a 'pointed expression of Rome's dominion over the earth and all her creatures'. But wild nature was plotting its revenge. Because Rome cast its nets so widely, it could not fail to haul in pathogens. Expanding to the border of the tropics, sending merchants to buy spices in the east, slashing forests, building roads through swamps, the empire created a novel 'disease ecology' where the microbes of the ancient world were pulled towards the centre, with its rats and flies and free rein for faecal-oral transmission. It was a petri dish for the first pandemics in history. Harper does not hide the connection to contemporary concerns: he writes with the TV on and recites Nipah, Ebola, SARS and MERS as correlates for what the Romans had to contend with. And the Romans also had a climate in flux. They built their civilisation in humidity, warmth and climatic stability, but halfway into it, these conditions were revoked. The agricultural lands feeding the empire dried out, some chilling and freezing; everywhere the predictable calendar of the seasons seemed gone. A millennium and a half before the steam-engine, this climate change was not anthropogenic – it originated in tiny shifts in the orbit of the earth and the output from the sun and a series of volcanic eruptions – although Harper opens for deforestation as a local driver, the clouds disappearing with the forests. The catastrophic effects were undiminished.

The dying of Rome, in Harper's telling, unfolded in several staggered steps. In the late second century, a wild germ from Africa made its way into the heartland and infected millions with fever, vomiting, lesions, rashes, bleeding: the Antonine Plague, pandemic No. 1, scything the population and leaving the empire down but not out. Rome soon rose again to new heights. Then, in the middle of the third century, withering droughts conspired with a second round of pestilence to blight the empire with twice the force. This time, the spillover might have come in the shape of a flu virus, possibly, Harper thinks, from migrating birds pushed by climate perturbations to fly over pigs and poultry. Now the energy reserves of the empire began to run low. In the late fourth century, a dust bowl in central Asia sent the Huns – 'armed climate refugees on horseback' – to drive the Goths towards the west, where the gates could not stand the pressure. The hordes entered Italy. In the early sixth century, the Plague of Justinian and the Late Antique Little Ice Age delivered the *coup de grâce* to the greatest civilisation the world had ever known, up to that point. In sum, 'the combination of plague and climate change sapped the strength of the empire'; for good measure, Harper adds the testimony of a survivor of the final carnage: 'The end of the world is no longer just predicted, but is revealing itself.' It's a few words of wisdom for the present. 'A precociously global world, where the revenge of nature begins to make itself felt, despite persistent illusions of control . . . this might feel not so unfamiliar', we read on the last page.

The fit is probably too good to be true. In the kind of debate that takes place at a remove from the reading public – similar to the aftermath of Jared Diamond's scientifically disreputable bestseller *Collapse* – the leading Marxist historian of late antiquity and climate change John Haldon teamed up with five other specialists to publish a rebuttal of *The Fate* in three instalments. On point after point, they catch Harper in the act of fiddling with the evidence. He cherry-picks anecdotes. He leaps to conclusions from one papyrus, uses sources from the wrong decade, misinterprets weather observations, frames the lack of evidence as proof of mass mortality, invents periods – anything to craft a juicy narrative, or so Haldon and colleagues claim. This is not the place to sift through the heaps of Roman ruins and assess the balance of evidence. But it appears that Harper is guilty of gross oversimplification, above all by eliding the central question of *how environmental stress factors were translated into hardship for the dominant classes* – he just assumes that this must have happened. Here, he echoes one streak in O'Connor, namely a degree of environmental determinism that takes for granted that ecological tribulation will fell the powers that be. But what must be examined is precisely the *mediation* of such disasters through relations between classes and states and other protagonists in a social formation. 'Simply assuming a causal connection is not sufficient', Haldon et al. admonish. Worse, simply assuming that ancient civilisations collapsed like ours seems about to do is positively misleading.

This has recently become a temptation for scholars in the field: to read current developments into what happened thousands of years ago. If a correlation between some variety of social breakdown and climatic fluctuation and (less commonly) infectious disease can be spotted, the seduced will announce causation. On a hunt for the cautionary tale, they will make out another *Titanic* in the mists of time and show it to the passengers of today. Hereby they construct a teleology, in which human civilisations have always had a tendency to self-destruct via the environment, only worsening over time; absent mediation, the picture is one of present-past symmetry. Disasters strike, civilisations fall. But *this time is qualitatively different*, for, among others, two elementary reasons.

First, episodes of non-anthropogenic climate change in human history cannot be compared to global heating, any more than a pub brawl to a genocide. Never before has our species stared at a breakneck destabilisation of the planetary climate system. Insofar as regional fluctuations did contribute to past collapses, these were different matters entirely. Low water levels in the Nile have been adduced as one cause of the very first such event, the falling apart of the Old Kingdom of Egypt; but they did not make life in that country impossible. On the contrary, lives of working people in Egypt by all accounts *improved* once the Pharaohs were gone, as the yoke of exploitation was lifted and the riverbanks could be cultivated with a touch of freedom. The fall of Rome might have been a similar godsend for the plebs, judging by how they threw themselves from the grip of Byzantium into the arms of Islam. The popular relief from ancient collapse seems to have been rule more than exception. It would, in other words, have been in the material interest of non-dominant classes to *let those empires self-destruct*, so that life could resume without bloodsuckers. No such silver lining exists today. The indications so far are that if *The Fate of Rome* scenario were to play out on a global scale (as Harper seems to expect), and if the second contradiction were to wear capitalism away from the planet, there would also remain precious little life worthy of the name. A hothouse earth on the way to Venus would be no place for a fresh start. The historical task must be one that no ancient populations ever faced: to consciously intervene so as to stop this civilisation from destroying itself by destroying the foundations on which any organised life must stand. The risk, clearly, is that a climatic second contradiction hits home after too much damage has already been done.

Secondly, Rome experienced no 1917. Neither did Egypt or Mesopotamia. To stay with Rome, it never experienced a revolution that broke off large chunks of territory *in a conscious attempt to transcend the imperial civilisation.* Nor was the agenda of the senate and the populace for a long time afterwards dominated by parties and movements from the same anti-Roman clade. The Forum did not reverberate with debates about how to achieve the aims of October in a better way, without the mistakes. Capitalist civilisation alone has produced such a legacy of immanent transcendence. It can now appear like another ancient Atlantis. If we date the chronic emergency to this millennium, organised anti-capitalism was mostly a spent historical force when it set in. But there is one possibility that cannot yet be written off: that some of that legacy will make a late comeback on the way down (never a possibility in Rome). Socialism is a seed bank for the chronic emergency. The anti-capitalist clade branched out in search for effective strategies of conscious intervention; a politics of conscious intervention is precisely what must now be revived. What can be of any use?

A brief obituary for social democracy

Our discussion must here be synoptic and syllogistic to the point of brutality. Social democracy as we now know it underwent its moment of speciation when Eduard Bernstein began to question the orthodoxy of revolution. His essential postulate was the absence of crises. The Steven Pinker of socialism, he pointed to the empirical fact that no serious crisis had rocked the capitalist economy for the past two or three decades, which invalidated the Marxian prophecy of a system trending towards collapse. Since it was not prone to malfunctioning, the idea of seizing power, smashing decrepit capitalism and installing a completely different order had become redundant; instead social democracy could continue to grow in strength, extract piecemeal reforms and gradually lift the working class out of the mire. Rosa Luxemburg very famously objected that the crisis tendencies had merely been postponed. In the near future, they would burst forth with even more dreadful violence. Ignoring her prognosis, the social democrats in the making went ahead and presently gave their first demonstration of how they dealt with catastrophe: by expediting it through consent. Since that moment of bifurcation, catastrophes have ever been the most inglorious occasions for social democrats. In their country of birth, they handled the second major disaster of the twentieth century – the rise of Nazism – in a similar fashion, tolerating every right-wing government up to the *Machtergreifung*, having no part in the destruction of the Third Reich. When social democracy did have its days of glory, it was in the most tranquil societies modern history has known. The best of the best cases might be Sweden from the 1950s to the mid-1970s, recently the object of retroactive adulation among young social democrats in the US. One thing is for sure: such calm days are not returning in the chronic emergency.

Social democracy works on the assumption that time is on our side. There must be plenty of it. Then one can move slowly towards the good society, step after incremental step, without having to clash head-on with the class enemy and break up its power; it will rather leak away in drips. But if catastrophe strikes, and if it is the status quo that produces it, then the reformist calendar is shredded. Social democracy can now do one of two things. It can continue to flow with the time, deeper into catastrophe – the choice from August 1914 – or it can become something else, another taxon of socialism, one that recognises that time is up and another decade or even year of this status quo is intolerable.

This is not to argue that actually existing social-democratic formations can have no role to play. To the contrary, they might be our best hope, as they have just been for a couple of years. Nothing could have been better for the planet than Jeremy Corbyn becoming the prime minister of the UK in 2019 and Bernie Sanders winning the presidency of the US in 2020. Had they been in charge of the two classic citadels of the capitalist core, there would have been some real opportunities to turn the present crisis and the others in the pipeline into ruptures with business-as-usual. What this means is merely that social-democratic politics in the chronic emergency *would have to go beyond itself* lest it relinquish the minimum aims – things like a tolerable life for all – rather as neoliberal governments have had to go beyond their own dogmas under the pressure of events. The time for gradualism is over.

A brief obituary for anarchism

Alpha and omega of anarchism: the state is the problem, statelessness the solution. After the Stalinist century ended with the whimper of 1989, this seemed like an appealing autopsy and cure-all to many (the present author included). Post–Berlin Wall social movements bent towards the libertarian branch of the old tree. Activists who came of age while internalising this hangover learnt to idealise the horizontal network, disdain leadership, spurn programmes, denounce the notion of seizing the state as the expressway to Gulag and embrace either anarchism wholesale or quasi-anarchist slogans such as 'change the world without taking power'. There was also a spurt of anarchist scholarship, one of whose main exponents is James C. Scott. He happens to be the one respected anarchist thinker with an interest in aspects of epidemiology (notably the insalubrious effects of the Neolithic revolution). The couplet state-as-problem / statelessness-as-solution runs through his oeuvre. What does it mean in practice? A true anarchist, Scott is loath to formulate political demands, but in *Two Cheers for Anarchism*, he invests a good deal of energy in at least one concrete proposal for making the world a better place: abolish traffic lights. If only the deadening hand of the state were removed from the streets, there would be more 'responsible driving and civic courtesy'. Anyone who has spent some time navigating streets in less affluent parts of cities like Tehran or Cairo would raise an eyebrow at that nostrum, but it faithfully reproduces the logic of anarchism: get the state off our backs. In the streets, in the workplaces, everywhere.

How do you apply this to something like corona? Or climate? The early weeks of the pandemic saw an efflorescence of 'mutual aid', as neighbours and local communities formed groups to help those in dire straits. They would pick up prescriptions or do the shopping for self-isolating elders, distribute masks, organise food banks, provide aid packages – toilet paper, bottled water, underwear – or produce hand sanitisers in a spirit of DIY, coordinating through Facebook or WhatsApp, very rarely appointing leaders. Some caught a glimpse of the mutual aid utopia long heralded by anarchists. In the favelas of Rio de Janeiro, another kind of community spontaneously stepped into the breach: drug traffickers, who announced and patrolled curfews to stop the virus from burning through the slums. 'The traffickers are doing this because the government is absent. The authorities are blind to us', one resident of Cidade de Deus told the *Guardian*. Hardly the ideal situation; more like a symptom of a deeply dysfunctional state beholden to the most depraved fractions of the dominant classes. Indeed, the mutual aid initiatives that made the greatest difference to the most vulnerable were probably those that took upon themselves tasks *the state ought to have shouldered* if only it hadn't been so lousy – a minus sign, not a plus. Communities under the auspices of states with reasonably well-functioning welfare sectors fared better than those that had to get by in a condition of statelessness. It is hard to see how any demand corresponding to the logic of abolishing traffic lights would help people in a pandemic. And then we are still stuck on the left side of the equation.

The case could be made that the second decade of the millennium was bookended by two events that disproved the alpha and omega of anarchism: the Arab Spring and Covid-19. In the most trenchant analysis of the former, *Revolution without Revolutionaries: Making Sense of the Arab Spring*, Asef Bayat shows how the key episode in particular, the Egyptian revolution, came to naught because of its complete lack of administrative authority. It never seized the state. Hence the state remained in the hands of retrograde forces, biding their time before they could annul every single gain from the fall of Mubarak – and part of the explanation was that activists wouldn't deign to storm the Winter Palace when the path to it lay open. They rose and fell with the anarchist zeitgeist.

> Revolutionaries remained outside the structures of power because they were not planning to take over the state; when, in the later stages, they realized that they needed to, they lacked the resources – unified organization, powerful leadership, strategic vision, and *some degree of hard power* – that would be necessary to wrest control both from the old regimes

and from other reactionaries, such as Islamists. Statelessness was not the solution in the Arab Spring. The state per se was not the problem in Covid-19. And if there is anything that will be needed on the right side of the equation in the chronic emergency, it is some degree of hard power.

The impending catastrophe and how to combat it

In the second week of September 1917, Lenin penned a long text called *The Impending Catastrophe and How to Combat It*. 'Unavoidable catastrophe is threatening Russia', it begins; the breath of death is over the land and 'everybody says this. Everybody admits it. Everybody has decided it is so. Yet nothing is being done.' World War I, the Ur-catastrophe of the century, had haemorrhaged Russia and the other belligerent countries and, so it seemed, put civilisation itself on the deathbed. 'The war has created such an immense crisis, has so strained the material and moral forces of the people, has dealt such blows at the entire modern social organisation, that humanity must now choose between perishing' or transitioning to 'a superior mode of production'. Russia stood before the spectre of famine. The war had so torn apart the country that all production apparatuses and logistical structures that would normally ensure basic provisioning were out of commission and, for as long as the war went on, beyond repair. As if that were not enough, heavy floods in the spring of 1917 washed away roads and railway lines. The crisis took a new plunge in August, when grain prices suddenly doubled and Petrograd faced the challenge of surviving without flour. 'Famine, genuine famine', one government official complained, 'has seized a series of towns and provinces – famines vividly expressed by an absolute insufficiency of objects of nutrition already leading to death'. It was in this situation that Lenin wrote his text. In the run-up to October, he and the Bolsheviks were suspended in a moment of abysmal emergency: war behind them, war to the side of them, famine advancing. Lenin obsessed over the breakdown. 'We are nearing ruin with increasing speed', he would write; 'no progress is being made, chaos is spreading irresistibly'; 'famine, accompanied by unprecedented catastrophe, is becoming a greater menace to the whole country week by week'. What could be done about it?

Part of the answer had already been provided by the states fighting the war. To prevent their food systems from collapsing utterly, they had interfered in markets in a manner that pre-war liberal doctrines would never have licensed. Governments from Paris to Petrograd had 'outlined, determined, applied and tested a *whole series* of control measures, which consist almost invariably in uniting the population and in setting up or encouraging unions' and rationing and regulating consumption. The situation had itself 'suggested the way out' by calling forth 'the most extreme practical measures; for *without* extreme measures, death – immediate and certain death from starvation – awaits millions of people'. But those measures had an obvious limitation: they dealt with symptoms. The drivers of catastrophe were left untouched. The inter-imperialist war and its primum mobile – simple ordinary capital accumulation – were kept going, leaving procurement systems on the edge or, as in Russia, over it. Here, then, was Lenin's wager: to take measures of the kind already instituted by the warring states, step them up a notch and deploy them *against the drivers of catastrophe.*

First was to end the war. Second was to get the grain supplies under control, seize stocks from rich landowners, nationalise banks and cartels, end private property in the key means of production – a revolution, as Lenin constantly agitated in these months, to stave off the worst catastrophe, which was why it must not be deferred. Against the Kerensky government's feeble attempts to restore order, he railed that 'it is *unable* to avoid collapse, because it is *impossible* to escape from the claws of the terrible monster of imperialist war and famine nurtured by world capitalism unless one renounces bourgeois relationships' and 'passes to revolutionary measures'. At the same time, his rhetorical gambit was to profess that the means for achieving this were at hand, almost uncontroversial. 'All the state would have to do would be to draw freely on the rich store of control measures which are already known and have been used in the past.' Indeed, he alleged that *any* government that wished to combat the impending catastrophe, whatever its affiliation, would have to take those radicalised measures. The objective logic of the situation left no other choice. Now, if we, for a moment, put aside the very considerable historical complications known to everyone, we can see that the logic of the present situation, *mutatis mutandis*, is not all that dissimilar. So what kind of control measures could be envisioned? Here we must again stay at the level of a rough sketch.

Yes, this enemy can be deadly, but it is also beatable

States in advanced capitalist countries could claim to have acted on the dangers of pandemics the moment they made the following announcement: today, we are launching a comprehensive audit of all supply chains and import flows running into our country. With our amazing capacity for surveillance and data collection, we'll shift from citizens to companies, open their books, conduct thorough input-output analyses (of the kind scientists already excel at) and ascertain just how much land from the tropics they appropriate. We shall then terminate such appropriation, by cutting off chains that run into tropical forests and, insofar as any can be classified as 'essential', redirect them to other locations. Every Noranda, every Skanska and Engie will be withdrawn. The time has come to pull in the claws of unequal exchange, now a menace to all.

We shall pay for tropical areas previously devoted to northern consumption to be reforested and rewilded. This will compensate for lost export revenues – not as charity or even a drain on our budgets, but as a running investment in the habitability of this planet, an establishment and maintenance of sanctuaries on which our health depends. We are here simply adhering to the categorical recommendations from scientists (whom we'll put on the stage for regular briefings on national television):

There is an urgent need to stop deforestation and invest in afforestation and reforestation globally. In response to the viral outbreaks, billions of dollars are spent on eradicating the infection, providing services to humans, and developing diagnostic, treatment and vaccination strategies. However, no or less attention is given to the primary level of prevention such as forestation and respecting wildlife habitats. The world should realize the importance of forests and the biodiversity carrying deadly viruses

– this from four China-based scientists, venting some despair amid Covid-19.

Similar advice has been given for years. 'The most effective way to prevent viral zoonosis is to maintain the barriers between natural reservoirs and human society.' Barriers? There is a force at work in human society that by its very nature cannot countenance such a thing. But again, the scientists: 'The most effective place to address such zoonotic threats is at the wildlife-human interface. A key challenge in doing this is to simultaneously protect wildlife and their habitats' – the most effective, and the most cost-efficient. 'Allocation of global resources from high-income countries to pandemic mitigation programs in the most high-risk EID [emerging infectious disease] hotspot countries should be an urgent priority for global health security', says the Pike paper. It estimates a tenfold return on such investment. Written six years before Covid-19, it speculates on the damage a zoonotic pandemic could wreak on the world economy and finds that mitigation at the source – reining in trade-driven plantations, livestock, timber, mining – would be a fantastically optimal way of saving money. This is evidently not a guarantee that it will happen. But the northern states of our fantasy have now committed themselves to reason and proclaim: *this is the right and necessary thing to do, for us and everyone else on this planet.* The immediate beneficiaries will be people living in or next to tropical forests, always first in line for spillover. But our control measures will also spare ourselves from living under this Damocles sword to the end of our days.

So the war on wild nature starts to wind down. This begins with a ban on importing meat from countries in or bordering on the tropics. Can there be anything more non-essential? And yet beef is, as we have seen, the one commodity most destructive to these wonderlands of biodiversity. Meat consumption in general is the surest way to waste land, and any extensive reforestation – combined with a protein-needy human population of ten billion or more – presupposes its reduction. Mandatory global veganism would probably be the endpoint most salutary for all. It would give some room back to wild nature and disengage the human economy from the pathogen pools; increased meat consumption is the fastest way to dive deeper. But as economies are currently operating, neither vegans nor vegetarians in the North go (as we often like to think) free of guilt: soybean, palm oil, coffee, chocolate flow as much, or even more, into our stomachs.

Control measures for addressing spillover should not follow dietary guidelines, but latitudinal gradients and ecological knowledge. Given what we know about bats, their habitats must have priority, be it steak or flapjacks that stream out of them.

Clearly it would be the state that would have to do this. No mutual aid group in Bristol could even hypothetically initiate a programme of this kind. 'We need (for a certain transitional period) a *state*. This is what distinguishes us from the anarchists', with Lenin – or with Wallace: 'In the face of the potential catastrophe, it would indeed seem most prudent to begin placing draconian restraints on existing plantation and animal monocultures, the driving forces behind present pandemic emergence.' Note the word 'draconian'. Progressives of all stripes might shudder at it, but they should return to the chapter on the working day in the first volume of *Capital* – the ten hours' day being the original victory of the proletariat, realised when enforcement finally became a little harsh, after all the laxities and prevarication of the early factory legislation. One doesn't curb capitalist exploitation by carrots.

Tropical forests have a recent counterpart to the ten hours' day: the tenure of Lula. Between 2004 and 2012, deforestation in the Brazilian Amazon underwent its most rapid reduction in modern times, all the more remarkable for running against the trends in the rest of Latin America and Southeast Asia. By what means did the Lula governments accomplish this? By turning some degree of hard power on land-hungry capital: expanding protected areas, registering land properties, monitoring rainforests via satellites, *enforcing* the forest code and actually punishing those responsible for illegal logging. In 2012, the rate of deforestation stood 84 per cent below its peak of eight years prior. The country that holds two million species, or one tenth of the earth's total, gave its forests a reprieve, slashing CO_2 emissions by some 40 per cent – perhaps the most impressive mitigation of zoonotic and climatic disaster on record. It didn't last, of course. 'Rosa Luxemburg has a great line about revolution being like a locomotive going uphill: if it's not kept moving, it slides back, and reaction wins. The same can be said of reform. Lula's two terms could have been a good first act in a transition toward something else; but there was no plan for a second act', as one scholar of Brazil has noted. Instead came the far right and the abolition of every traffic light ever installed in the Amazon. What should really make one shudder is to think of the zoonotic and climatic legacy of Bolsonaro.

Then what of China? After SARS, the state took some perfunctory measures to stem the wildlife trade, promulgating laws with loopholes big enough for rhinoceroses to walk through. It allowed for wild animals to be bred on farms (the Huanong Brothers). The protected species list was last updated in 1990 and omitted at least one thousand native species – including bats – the consumption of which was thereby unregulated, regardless of the public health consequences. Penalties were paltry, enforcement

lax, 'high profits and mild punishment driving the dealers'
to continue accumulating capital – until SARS-CoV-2
prodded the state to ban the consumption of any wildlife,
from freedom or captivity. Scientists and others worried
that the legislation would fray this time too. One team
from China writing in *Science* urged a permanent ban on
consumption as well as possession, backed up by stiff
penalties; Jingjing Yuan and colleagues went a step further
and called for 'sentence to life prison' for anyone eating
wild. Processing, transporting, marketing wild animals
should be similarly sanctioned, the state maintaining a list
of species authorised for trade – a list that could be periodically shortened – and sending inspectors into the
markets on the fly (recalling the factory inspectors).

What could be said against such a tough line? It has
been argued that the moral norms of consumers should
instead be coaxed into sobriety. The argument ignores
three factors. First, if SARS was not enough to scare the
clientele away from wet markets – research indicates that
awareness of the risks did little to put it off – and if SARS-
CoV-2 could not be relied on to do the job either, as some
signs suggested – online sellers touted medicines contain-
ing rhino horn and other rare animal parts as *cures* for
corona – then apparently one cannot entrust this question
to individual enlightenment. Second, enforced laws
change norms. The prohibition of child labour in factories
and slave labour on plantations clinched their status as
unacceptable practices; without those laws, some callous
exploiters might have continued to this day. The

edification may outlast the laws themselves. One of the few success stories Felbab-Brown can relate in *The Extinction Market* concerns the use of rhino horn for the making of the Yemeni daggers known as *jambiyas*. When demand soared in the 1970s, this market became a prime culprit in dragging rhino populations to extinction. But then someone intervened.

> Interestingly enough, the communist government of South Yemen was far more effective in eliminating demand for rhino-horn *jambiyas* by eliminating the demand for all *jambiyas*. It banned the possession of all weapons and aggressively collected them. In 1972, the *jambiya* ban was thus accompanied by a massive campaign to rid the country of them, with even rich and influential families targeted and forced to sell their daggers.

When Yemen was reunited under the capitalist north, the communist principle survived. The ban 'was not only effectively enforced by the [southern] government but ultimately internalized by the country's population'. Rhino-horn *jambiya* went out of fashion. This deep into the sixth mass extinction, some similar courage to wage ecological class war would not seem inappropriate.

Third, if there is something the corona crisis has taught, it should be that nudging consumers to voluntarily mend their ways is a strategy of the past. The German state didn't beg its citizens to please consider living differently: it ordered the malls of Steglitz closed and locked the playgrounds in Kreuzberg. When there is a threat to the health or even physical existence of a population, one doesn't leave it to the least conscientious individuals to play with the fire as they want. One snatches the matches out of their hands. Some have argued that a blanket abolition of the wildlife trade in China would cause financial losses and make people unemployed – figures between 1 million and an improbable 14 million have been floated – which is, of course, the excuse for every facet of business-as-usual. It could carry us all the way to Venus.

But ending the wildlife trade is a responsibility for very many more nations than China. Even Germany has been identified as a central transit point for the global trade in pangolins. States have to figure out a way to extirpate this department of capital accumulation *in toto*; they have repressive powers to reallocate. Barack Obama purported to make crackdowns on wildlife trafficking a priority. Yet at the end of his second term, there were no more than 130 federal wildlife inspectors in the nation; only 38 of 328 ports of entry had such staff on site; their total number of detector dogs amounted to three. Compare this – from benevolent times – to the apparatus for stopping migrants. Here's another overdue conversion: open borders to people and close them to commodities from the wild; turn ICE and Frontex and other fortress guards into agencies for shutting down the extinction vortexes. But law enforcement would require more than seizures on the border, which can incite suppliers to compensatory killing sprees. It is the middlemen that need to be netted *en bloc*.

The main alternative to such an approach is to legalise the wildlife trade and encourage the ordered establishment of farms (the Huanong Brothers), but the curtain should now be down on this idea. Wild animals shouldn't sit in cages. Breeding them in captivity and selling them on markets only whets the appetite for their meat, and experience shows that it's all but impossible to tell the wild from the farmed; the former leaks into the latter, as long as the suck is there. Demand itself will have to be neutralised. Insofar as ostentation – the open display of status before peers and subalterns – is the purpose of wildlife consumption, criminalisation and actual law enforcement should hit where it hurts. Under the ground, public swagger is harder. This doesn't mean, as Felbab-Brown is keen to stress, that hard state power is a silver bullet. But it is needed, and fast, she points out. 'Unlike in the case of drugs' – and most other illicit activities, one may add – 'time matters acutely, especially when animals are being poached at extinction rates.' Some reprioritisation is needed for repressive state apparatuses around the world.

And then there is the question of bushmeat, an especially difficult nut to crack, which deserves its own separate investigations. One would wish that lifting areas and countries out of poverty would of itself make bushmeat obsolete, but alas, it might have the opposite effect: affluence can set the extinction vortex spinning. It has, on the other hand, been vociferously argued that one shouldn't even consider taking the wild food out of the mouth of poor people. Unfortunately, that argument is self-defeating, for in the same moment bushmeat starts to endanger animal populations, it ceases to be a prop of food security and turns into its opposite: an exceedingly undependable protein source. Extinction exhausts it forever. The most viable palette of measures probably includes laws and their enforcement, a rollback of defor-estation and 'incentives for communities to switch to traditionally grown protein-rich plant foods', such as 'soy, pulses, cereals and tubers' – breaking, in other words, the association of meat with the good life. That break begins in the richest countries. If anyone has a duty to lead and assist a global turn to plant-based protein, it is them.

Needless to say, such measures would just be starters – local drivers of deforestation, for instance, would still have to be dealt with – and if they were all rolled out next week, infectious diseases wouldn't thereby vanish at the snap of a finger. The treatment of symptoms will never stop being essential. And so one could look to Cuba, which seems to have spare capacity for every eventuality and continues to serve the world as a subaltern ambulance crew, including in this pandemic: in March 2020, fifty-three professionals in a Cuban medical brigade landed in Lombardy. They came to assist the swamped hospitals of one of the richest provinces in Europe. Of the dozen brigades dispatched over that month, others went to Jamaica, Grenada, Suriname, Nicaragua, Andorra, while

Cuba itself agreed to receive a corona-stricken cruise ship turned away from other Caribbean islands – all in line with a tradition of 'medical internationalism' that never ceases to confound foes and experts alike. In the 2010s, this poor little nation had more health care workers stationed on foreign soil than the G8; more than the Red Cross, Médecins Sans Frontières and UNICEF combined. When Ebola lacerated West Africa in 2014, hundreds of doctors and nurses dashed off to the miasmic front lines; when Hurricane Mitch tore through Central America and Haiti in 1999, not only did Cuban staff pour in, but Havana initiated a scholarship programme for medical students from the disaster zones; when an earthquake crushed Pakistan in 2005, Cuba sent 1,285 health workers for a year. Canada sent six. In a time of chronic emergency, the world should thank its lucky star there's at least one state with a tenuous link to the communist ideal still around.

War on the oil barons

'If anything real is to be done, bureaucracy must be abandoned for democracy, and in a truly revolutionary way, i.e. war must be declared on the oil barons and shareholders': Lenin. His *casus belli* was their refusal to produce enough oil and coal. He wanted a war on the barons and shareholders to force the pace of extraction – Russia 'is one of the richest countries in the world in deposits of liquid fuel' – having no inkling of any adverse effects. Fuel scarcity was part of his breakdown. Our breakdown has the opposite profile, and so, if anything real is to be done, there will have to be a war with another aim: putting this industry out of business for good. This begins with a nationalisation of all private companies extracting and processing and distributing fossil fuels. Corporations on the loose like ExxonMobil, BP, Shell, RWE, Lundin Energy and the rest of the pack will have to be reined in, and the safest way to do that is to put them under public ownership, either through acquisition or – more defensibly – confiscation without recompense. Then their endlessly burning furnaces can finally be switched off. But they should not simply be liquidated, as in dismantling every platform, sealing the holes, closing the offices, sacking the employees and throwing the lot of the technology on the scrap heap. To the contrary, these units have a constructive task ahead of them.

It's already too hot on earth, and it's getting hotter by the year, and there's no end in sight to the heating unless emissions are cut to zero – but even then, it will still be too hot plus residual, potentially self-reinforcing heating in the atmospheric pipeline (the more of it, the longer mitigation waits), and so a worldwide cessation of fossil fuel combustion would not be enough. CO_2 would also have to be drawn out of the air. This has been apparent for at least a decade: everybody says this. Everybody admits it. Everybody has decided it is so. Yet nothing is being done. Nothing at all? There are a bunch of start-ups developing machines for negative emissions. One of them, the

Swiss-based Climeworks, might be the most valuable capitalist company on earth these days – valuable as in doing humanity what could eventually be a life-saving service.

With machines that look like large fans in boxes, Climeworks sucks air – it could be any air, anywhere. The air is led into a filter that captures CO_2. Once the filter is saturated, it is heated to 100 degrees Celsius, and the result is pure, concentrated carbon dioxide. The trick as such is no magic, as it has long been applied in airtight rooms – submarines, space stations – where CO_2 has to be scrubbed and flushed out for people to breathe. What Climeworks has just demonstrated, however, is that this is the most promising technology for taking CO_2 out of the earth's atmosphere – far more so than 'bioenergy carbon capture and storage', or BECCS, the speculative solution most in vogue in the days of the Paris agreement. There the idea was to establish gargantuan plantations to culti-vate fast-growing trees, harvest them, burn them as fuel, filter away the CO_2 and store it under the ground. But more plantations are not what we need. BECCS would devour such monstrous amounts of land – somewhere like the equivalent of *all current cropland* to stay below 2°C – that tropical forests might well have to be wiped out. *Direct* air capture needs no land to grow anything. The contraptions can be placed on roofs. The main inputs they crave are electricity and heat, and because they are small and easily switched on and off, they can be affixed to the grid and turned on when there is an excess of wind and sun (weather-determined moments of overproduction often regarded as a drawback of renewables) and use waste heat from any other process (no shortage of that in urban environs). The CO_2 can be mineralised. It can be buried under the ground in solid form; indeed, since 2017, Climeworks is doing just this in Iceland. As with other novel technologies – solar panels spring to mind – prices will nosedive with mass production.

A capitalist solution to a problem made by capitalism? If only. A capitalist company has to have a commodity to sell. With the exception of the pilot plant in Iceland, Climeworks and the other start-ups are turning their concentrated CO_2 into goods with exchange-value. It can be gas sold to greenhouses or soft drink producers (Coca-Cola in the case of Climeworks in Zürich); it could go into microalgae or liquid fuel, possibly even for airplanes. Such commodities bury no CO_2. They capture it and pass it on for release elsewhere, so that a profit can be made – or, as *Nature* reported regarding another start-up, Carbon Engineering, run by the famed scientist-cum-entrepreneur David Keith: 'That CO_2 could then be pressurized, put into a pipeline and disposed of under-ground, but the company is planning instead to use it to make synthetic, low carbon fuels.' And how could it plan otherwise? Just throwing the CO_2 away, locking it up in cellars where it must never again be touched, is no way to accumulate capital. It negates the logic of the commodity, because non-consumption would here be the innermost essence of the operation. As Holly Jean Buck shows in

After Geoengineering: Climate Tragedy, Repair, and Restoration, a primer and clarion call that should be obligatory reading for anyone minimally concerned with planetary futures, this is the contradiction every direct air capture must run into: if it stays inside the commodity form, it cannot make good on its promise of negative emissions. It will recycle CO_2, not tuck it away.

To scale up these machines to the level where they would make their designated difference – supplementing zero emissions with drawdown – they would have to function as vacuum cleaners, sucking up carbon and putting it *out* of circulation, as a non- or even anti-commodity. How could such a decontamination of the biosphere run on profit? Where would the increment in exchange-value come from, in amounts sufficient to keep the clean-up going like any other department of accumulation? No one has yet come up with a plausible answer. Buck works through the logic and finds only one way out: the state. Other students of direct air capture have reached the same conclusion. It seems to inhere in it – if the Climeworks model turns out to have some unknown disadvantage, if something else comes to the fore as the superior tech, if there will ever be any negative emissions not growing from land, the same conundrum will reappear: resell the waste and forfeit the purpose, or respect the negative use-value. It's the productive force *or* the property relations.

And to scale up, one would need a lot of money. That money should come from those who carry historical responsibility for releasing the CO_2 in the first place.

There would also need to be massive complexes of technical expertise, drilling and seismic skills, infrastructures for transporting concentrated CO_2, empty holes in the ground for burial vaults, organisations of supranational size . . . Who has all these things in ample possession? The oil barons and shareholders, of course. Nationalise them, Buck proposes – not just for 'getting rid of these corporations, as we might like to, but transforming them into companies that deliver a carbon removal service'. Make them public utilities for restabilising climate. In something of an understatement, Buck adds: 'There will be a lot of struggles to engage in here.'

But now imagine that states were in fact determined not only to stop the drivers of catastrophe but to put them into reverse gear, and so they expropriated every single fossil fuel company and restructured them into waste disposers, while those already state-owned received the same directives – then we would really be on the way to zero emissions and further: towards 400 parts per million, 380, 350 . . . It would be some repair to match the tropical rewilding. The demand for nationalising fossil fuel companies and turning them into direct air capture utilities should be *the* central transitional demand for the coming years. But, needless to say, it would make no sense if CO_2 were still belching out into the atmosphere: emitting *and* capturing would be a bizarre dissipation of resources to no avail. Everything begins with draconian restraints and cuts. They alone could pave the way for actual drawdown; the sooner they start, the less need for a secondary mega-infrastructure of clean-up.

The problem could also be attacked from another angle: not supply but demand, rather like in the first phase of the Covid-19 pandemic. Then it was demand, above all in the transport sector, that went off a cliff and pulled emissions along. In late April 2020, *Scientific American* publicised the forecast that total global emissions would fall by no more than 5 per cent during the year – in spite of the spring drop by one fourth in China and roughly one fifth in the US – as economies were expected to rebound in the summer and autumn. The journal noted that as record-breaking as a 5 per cent reduction would be, it would still fall short of 'the 7.6 per cent decline that scientists say is needed every year over the next decade to stop global temperatures from rising more than 1.5 degrees Celsius'. Nearly 8 per cent *every year* – a far cry indeed from the expected 2020 hiatus (if not from the initial months-long collapses). What would that require? Comprehensive, airtight planning. Everybody knows this. Few say it. One can obviously not rely on spontaneous cessation of demand, or on people just quitting travel; there would have to be a continuous substitution of one kind of energy for another over the transitional period – or, 'a single economic plan covering the whole country and all branches of productive activity. This plan must be drawn up for a number of years, for the whole epoch that lies before us', to cite Leon Trotsky. One can of course find this idea so repugnant that one would rather give up on the climate of the earth. And that is indeed the choice the dominant classes and their governments wake up to make every morning.

Regardless of whether the problem is attacked from the supply or the demand side, the race to zero would have to be coordinated through control measures – rationing, reallocating, requisitioning, sanctioning, ordering . . . – so as to fill the gap after fossil fuels. The substitutes themselves are in no need of elaboration. The literature on the Green New Deal and renewable energy roll-out and climate wartime mobilisation is extensive enough to guide a transition several times over. Here we truly are in the situation of Lenin's September text: everybody knows what measures need to be taken; everybody knows, on some level of their consciousness, that flights inside continents should stay grounded, private jets banned, cruise ships safely dismantled, turbines and panels mass produced – there's a whole auto industry waiting for the order – subways and bus lines expanded, high-speed rail lines built, old houses refurbished and all the magnificent rest. 'The ways of combating catastrophe and famine are available', approaching common knowledge. 'If our state really *wanted* to exercise control in a business-like and earnest fashion, if its institutions had not condemned themselves to "complete inactivity" by their servility to the capitalists, all the state would have to do' would be to roll up the sleeves. Another part of Lenin's logic applies too: *any* government that would 'wish to save Russia from war and famine' would have to get down to this kind of work.

But the lingering conclusion from our initial comparison between corona and climate is that *no capitalist state is likely ever to do anything like this* of its own accord. It would have to be forced into doing it, through application of the whole spectrum of popular leverage, from electoral campaigns to mass sabotage. Left to its own devices, the capitalist state will continue to attend to symptoms, which, however, must eventually reach a boiling point. One can imagine that in the next years and decades, storms will bite into property, droughts tear apart supply chains, crop yields halve, heat waves enervate labour productivity to the extent that the timeline of victimhood catches up with the dominant classes. The second contradiction will then be upon them. States might no longer be able to just parry the impacts, but feel compelled to safeguard the background condition before it crashes irretrievably. Judging from the reaction to Covid-19, they will grasp for a control measure that can flatten the curve *at once*, and there is one such known in the libraries of science: solar geoengineering. Spraying sulphate aerosols into the atmosphere is the single kind of injection with a potential to instantly reduce planetary fever. However large in scale, direct air capture would need decades to bring temperatures down; sulphate aerosol injection can cut insolation from one month to the next. Year after year of business-as-usual, this is the pseudo-solution that sneaks up on us like a thief in the night.

Indeed, under the cover of the pandemic, in mid-April 2020, one of the largest experiments in geoengineering so far was carried out on the Great Barrier Reef, then subject to the third outbreak of mass bleaching in five years (did anyone notice?). Scientists were authorised by the state to spray trillions of nano-sized ocean salt crystals into the air from the back of a barge. The hope was that these particles would make clouds brighter, so they would reflect more sunlight away from the ocean and shield the reef from the heat. The team told the *Guardian* they could see corals 'bleaching around us' as they bobbed over them. This is a technology distinct from sulphate aerosol injection, namely marine cloud brightening, potentially deployed on a local or regional scale by a state such as Australia, which, numerous monumental disasters notwithstanding, cannot bring itself to impose any control measures on coal extraction. The logic is robust. As one of the sharpest scholars in the field, Kevin Surprise, has argued, solar geoengineering might well be launched on a planetary scale as a fix against the second contradiction, because capitalist states appear constitutionally incapable of going after the drivers. It is fairly widely known that such intervention in the climate system could switch the planet onto another track towards catastrophe. Meanwhile, the corals keep bleaching, the swarms forming, the ice melting, the animals moving.

A pestilential breath devastating humanity

There has been a lot of talk about ecological Marxism in recent years, and with the chronic emergency over us, the time has come to also experiment with ecological Leninism. Three principles of that project seem decisive. First, and above all, ecological Leninism means *turning the crises of symptoms into crises of the causes*. From August 1914, this was, of course, the thrust of Leninist politics: converting the outbreak of war into a blow against the system that engendered it. Our Great War is not an actual war between armies, nor a singular event that can be concluded or paused after half a decade: this emergency is chronic, which means that crises of symptoms will ignite *again and again*, and every time they do, the strategic imperative must be to switch energies of the highest voltage against the drivers. It is difficult to see how else the conditions can ever be ameliorated. Has anybody got another idea? Oh yes: make clouds and invent vaccines; block solar radiation and track the movements of people. At their best, such proposals amount – to borrow from Greta Thunberg's favourite metaphor – to surviving inside a burning house by drinking lots of cold water. Virtually by definition, the most classical Leninist gesture is the only one that can point to an emergency exit.

It is worth re-emphasising just how central the category of catastrophe was to the evolution of revolutionary Marxism. In her polemics with Bernstein, Luxemburg never tired of stressing it. She has become most renowned for the sound bite 'socialism or barbarism' but, as Norman Geras has shown in a superb exegesis, that deep dichotomy structured her theory and praxis all the way from the battle with Bernstein to her death at the hands of the

Freikorps. One year into the war, she warned that human-ity faced a choice between 'the destruction of all culture, and, as in ancient Rome, depopulation, desolation, degen-eration, a vast cemetery' – or victory for 'the conscious struggle' against the imperialism that drove the war. 'Wading in blood and dripping in filth', capitalist society has become 'a pestilential breath, devastating culture and humanity'. That peculiar type of society now 'endangers the very existence of society itself, by assembling a chain of devastating economic and political catastrophes'; in its present phase, the expansion inherent in capital 'has adopted such an unbridled character that it puts the whole civilisation of mankind in question'.

Luxemburg expected world war to become a 'perma-nent' state of affairs. It didn't, and here the *differentia specifica* of the chronic emergency must again be under-scored: it works itself out through biophysical processes that cannot be fought or negotiated to an end. One does not bomb out or bargain with the radiative forcing of CO_2. That forcing is an immutable function of the quan-tity of the gas in the atmosphere, which means that this pestilential breath has another order of permanency and aggravation – until the moment of deliberate intercession, still only hypothetical. Following Geras's reading of Luxemburg, we can then say that 'barbarism', depopula-tion, a vast cemetery really are the inevitable ends of a capitalism left to itself (here precluding the long-term effectiveness of solar geoengineering as a stand-alone measure). But writing in 1975, he recoiled from this conclusion as excessively apocalyptic. 'Ecological catastrophe may, today, be invoked to lend that vision plausibility', he noted in passing; half a century later, there is scant need for the caveat.

This, then, is the syntax of revolutionary Marxism, present already in the first section of *The Communist Manifesto*: the fight ends 'either in a revolutionary reconstitution of society at large, or in the common ruin of the contending classes'. There can be little doubt about which of the two outcomes is currently the more likely. Hence the accentuated 'conditional mood of the probability of a catastrophe that there is still time to forestall. Things will end up badly, if . . . But they can (still) be sorted out . . .', as another thinker from the same tradition, Daniel Bensaïd, distils the predicament. It was because Luxemburg threw herself into efforts to forestall further catastrophe that she, for all their disagreements, ended up on the same side as Lenin. A second principle for ecological Leninism can be extracted from their position: speed as paramount virtue. 'Whether the probable disaster can be avoided depends on an acute sense of conjuncture', writes Bensaïd, who reconstructs the crisis of September and observes that 'waiting was becoming a crime'. Or, with Lenin himself: 'delay is fatal'. It is necessary to act 'this very evening, this very night'. The truth of these assertions has never been more patent. As anyone with the barest insight into the state of the planet knows, speed, very regrettably, because of the criminal waiting and delaying and dithering and denying of the dominant classes, has become a metric of meaning in politics. 'Nothing can now be saved by half-measures.'

Third, ecological Leninism leaps at any opportunity to wrest the state in this direction, break with business-as-usual as sharply as required and subject the regions of the economy working towards catastrophe to direct public control. It would mean that 'one part of the population imposes its will upon the other part', to speak with Engels. Nothing from the past decades of stalled transitions indicates that ExxonMobil would like to metamorphose into a cleaner and storekeeper of unsalable carbon, or that meat and palm oil companies would gladly let their pastures and plantations be rewilded. It appears tautologically true that an actual transition would require some coercive authority. If anarchists would ever wield influence in such a process, they would quickly discover this circumstance and, just like anybody else, have to avail themselves of the state.

But what state? We have just argued that the capitalist state is constitutionally incapable of taking these steps. And yet there is no other form of state on offer. No workers' state based on soviets will be miraculously born in the night. No dual power of the democratic organs of the proletariat seems likely to materialise anytime soon, if ever. Waiting for it would be both delusional and criminal, and so all we have to work with is the dreary bourgeois state, tethered to the circuits of capital as always. There would have to be popular pressure brought to bear on it, shifting the balance of forces condensed in it, forcing apparatuses to cut the tethers and begin to move, using the plurality of methods already hinted at (some further outlined by the present author in *How to Blow Up a Pipeline: Learning to Fight in a World on Fire*). But this would clearly be a departure from the classical programme of demolishing the state and building another – one of several elements of Leninism that seem ripe (or overripe) for their own obituaries.

On the other hand, the chronic emergency can be expected to usher in pronounced political volatility. 'The deeper the crisis, the more strata of society it involves, the more varied are the instinctive movements which criss-cross in it, and the more confused and changeable will be the relationship of forces', to quote Georg Lukács. The rather startling measures used to combat the spread of Covid-19 might have been a foretaste. Who knows what openings other moments of impact might bring. In some, popular initiatives may rise to prominence. The 2013 edition of the 'worldwide threat assessment' compiled by the US intelligence community warned that climate disasters risk 'triggering riots, civil disobedience, and vandalism'; similar predictions are legion. If or when they are fulfilled, the mission of ecological Leninists is to raise consciousness in such spontaneous movements and reroute them towards the drivers of catastrophe. Hence the heightened relevance of the slogan that for Bensaïd 'sums up Leninist politics: "Be ready!" Be ready for the improbable, for the unexpected, for what happens.' It includes a readiness to, with Lenin's own words, 'set to work to stir up all and sundry, even the oldest, mustiest and seemingly hopeless spheres, for otherwise we shall not be able to cope with our tasks'. If the matter is exigent, the material at hand must be used.

On this view, ecological Leninism is a lodestar of principles, not a party affiliation. It does not imply that there are any actual Leninist formations capable of seizing power and implementing the correct measures – the world has never been shorter on them, and most of the few that remain show overt signs of infirmity. The old Trotskyist formula 'the crisis of humanity is the crisis of the revolutionary leadership' must be updated. The crisis is the absence – the complete, gaping absence – of any leadership. The seed bank exists in an arid space approaching empty desert; anything brought out from it would have to be genetically modified to grow under the present sun and watered by subjects inventing themselves anew. Two elements do, however, as we have argued, appear essential. The basic make-up must harbour a predisposition for emergency action and an openness to some degree of hard power from the state. Anarchism detests the state; social democracy shrivels in catastrophe. But there is no reason not to experiment with ecological Luxemburgism, or ecological Blanquism, or Guevarism, or indeed Trotskyism . . . nor is there reason to give up on the sheer deductive force of revolutionary Marxism: 'The inherent tendencies of capitalist development, at a certain point of their maturity, necessitate the transition to a planful mode of production, consciously organised by the entire working force of society – in order that all of society and human civilisation might not perish', again with Luxemburg. But 'necessitate' does not mean 'preordain'. Something can be necessary and yet never come about.

Red Army of renewable energy

When Emmanuel Macron declared that 'we are at war' against Covid-19 and Bernie Sanders called for the US to approach climate change 'as if we were at war', the war in their minds was metaphorical. Macron didn't call up his soldiers to shoot at the virus. But metaphors are not innocent, and some have objected to the verbal belligerence, including on the climate front; two energy scholars, Johannes Kester and Benjamin Sovacool, have taken Delina and others to task for purveying an atmosphere of bloodshed. Peace is a better thing than war, they incontrovertibly argue. Hence the metaphor of peace ought to replace that of war. On this line of reasoning, there should be no talk of 'war on want' or 'class war' or 'climate wartime mobilisation', but rather 'peace on want' and 'class peace' and 'climate peacetime mobilisation'. Macron should have said 'we are at peace with Covid-19' and Hillary Clinton promised to install a peace negotiating table for climate change, perhaps at Camp David itself. What signals would that have sent? Metaphors of war roll off the tongue *when life and death on a mass scale are at stake* and *the situation demands extraordinary mobilisation to survive* and such episodes do transpire every now and then in human history. Then the language of peace doesn't work, for the simple reason that it conveys a state of bliss.

It wouldn't have worked much better if Bill de Blasio had said 'we're waging nonviolent civil disobedience against Covid-19 and ventilators are our flowers and songs'; it would barely have been intelligible. There is a reason Jesus himself announced 'I did not come to bring peace, but a sword', even though, in the extant accounts, he never brandished an actual sword. Bill McKibben is a Christian pacifist, and yet he does not shy away from saying, as we have seen, that global heating is not '*like* a world war. It *is* a world war', which is, of course, but another way to enhance the metaphor; he doesn't really think it is an intense armed conflict. 'War' signifies the arrival of catastrophe. The proliferation of martial metaphors in this historical moment is a deplorable but apt reflection of how things stand. Wishing it away doesn't help. The question is rather, as Alexandria Ocasio-Cortez pointed out in early January 2020, when the Australian state sent soldiers to evacuate citizens from the atomic bomb of the bushfires – the metaphor coming closer to literality – if we fight back against the symptoms or the drivers: 'will we mobilize reactively to destruction, or will we act proactively for prevention?' The former can only end in dire, terminal loss.

A second, more specific objection to World War II as metaphor and analogue says that the climate crisis is not as tangible as aerial bombardment, nothing comparable to Pearl Harbor; that we can disregard. Kester and Sovacool also claim that people will be scared away from climate action if it is framed in warlike terms. That is gainsaid not only by the general willingness to join the 'war' on Covid-19, but by psychological research showing that ordinary Americans consistently report greater sense of urgency and will to mitigation after being fed information about the climate crisis as a war, rather than as a 'race' or a non-metaphorical 'issue'. From this critique, World War II emerges unscathed as a point of reference, most useful for the speedy, state-led conversion of production. But it has other limitations. From the American perspective – as distinct from, say, the partisans, but they do not appear in this discourse – that war was fought for the sake of the status quo. It aimed to defend US capitalism as it was. Parts of the dominant classes might have felt inconvenienced by the temporary subjugation to military planning, but no fraction of them had to be abolished or humbled and pressed into other material moulds; they went along with the mobilisation. As for fossil fuels, the home front burnt as much coal as possible in the effort. The allied armies were giant moving pyres of oil. For these three reasons – it did not reorder the economy or confront vested interests, but deepened the reliance on fossil fuels – World War II as fought by the US is, Kester and Sovacool rightly argue, imperfect for the purposes of the present. It should not therefore be exchanged for wishful talk of peace. It should rather be complemented with another metaphor and analogue: war communism.

Hardly had the Bolsheviks stormed the Winter Palace, seized power, declared peace and exited the world war before a new war was thrown at them, even more savage and disastrous for Russia. The dominant classes swept away with minimal bloodletting in October regrouped and attacked from the west, north, south and east. United by revanchist royalism and ferocious anti-Semitism, the Whites were abetted by troops from the US, the UK, Germany, France, Italy, Japan, Canada and half a dozen other nations. All the worst-case scenarios from Lenin's *Impending Catastrophe* came to pass. Famine marched through the land. Snapping under the pressures of world war and civil war (of the fully internationalised kind), the food system failed to supply a minimum of nourishment to many millions; compounding its drawn-out collapse, a severe drought set off the worst sequence of starvation in 1920. Epidemics flew like swarms of bombers across Russia: there were outbreaks of typhus, cholera, smallpox, dysentery, bubonic plague and, of course, the Spanish flu, introduced by the American expedition landing in Arkhangelsk and spread further by the Japanese Imperial Army advancing through Siberia. 'Either the lice will defeat socialism or socialism will defeat the lice', muttered Lenin in December 1919. Victor Serge described how sewage pipes froze up inside city buildings. Desperate families would pilfer wooden furniture from mansions to throw on their fires; there was little other fuel to be had. 'Through the interminable nights of the Russian winter all lights were kept down to a flicker.'

The Bolsheviks, in other words, stumbled from one emergency into another, which spurred them to radicalise some of their policies. In the half-year before the Civil War, their government went slow on the nationalisation of enterprises. The process was instead rammed through from below, by local soviets and trade unions taking over factories, mines and workshops and then demanding support from the state; but once it became encircled by enemies, the state itself took to the offensive and concentrated the means of production in its hands. A slew of other control measures followed. Their emergency character was reflected in the irony that the Whites, fully committed to private property, also instituted forms of state control over their economies – including nationalising coal mines – to cope with the situation. Both took leaves from the war economies of Imperial Germany and Tsarist Russia: when in exigency, forget about the sanctity of property. But under Bolshevik rule, of course, the change went much deeper than anywhere else. In *Russia in Revolution: An Empire in Crisis, 1890–1928*, the most nuanced account to emerge from the centenary of October, S. A. Smith marvels at 'the incredibly rapid way in which the privileged elites disappeared. The major assets of the nobility were, obviously, taken when peasants seized landed estates; and capitalists lost their assets with the nationalization of industry, commerce, and banks.' For Smith, this evaporation of the dominant classes was an outstanding feature of what has become known as 'war communism'.

Now this term tends to leave an acid taste. Rightly so. The warring Bolsheviks committed no small amount of cruelty. Let it be said, then, that invoking war communism is not to suggest that we should have summary executions, send food detachments into the countryside or militarise labour, just as no one who looks at World War II as a model for climate mobilisation wants to drop another atomic bomb on Hiroshima. Many of the perceived necessities the Bolsheviks turned into virtues, we can readily recognise as vices. But, conversely, some of what they saw as their weaknesses we may regard as strengths. In *The Economic Organization of War Communism 1918–1921*, the weightiest work on the subject, Silvana Malle notes that the territory under Bolshevik control by late 1918 had shrunk to one-eleventh of that guaranteed by the peace treaty of Brest-Litovsk (already punitive to Russia). Gone were the coal mines of the Donets basin and the oilfields of Baku. White and allied forces occupied 99 per cent of the coal and 97 per cent of the oil resources on which the Russian economy had hitherto been based. The foreign blockade cut off imports; for all practical purposes, Bolshevik Russia was now a land without fossil fuels. Trotsky wrung his hands over the hopelessness of the situation. The right property relations mattered little if there was no energy to burn: 'An industry which is completely deprived of fuel and raw materials – whether that industry belongs to a capitalist trust or to the labour state, whether its factories be socialized or not – its chimneys will not smoke in either case without coal or oil.' And for a good two years, no red chimneys sent out such smoke.

What did the Bolsheviks do without fossil fuels? They turned to wood from the boreal forests of what remained of their Russia. This poor substitute had provided 17 per cent of the energy consumed by industry in 1913; seven years later, the share stood at 83. It might have been even larger in the critical sector of transport – 'we had to stoke our boilers with recently stored raw wood', complained Trotsky, train-rider-in-command. Malle describes how the state set up a 'supreme organ of fuel policy' to save the day. It carried out an inventory of stocks of timber, sent agents to verify local needs, assigned priority to certain sectors (notably the railways), introduced premiums to speed up collection and organised a nationwide system that made up for much of the shortfall and effectively, under the gun, turned Bolshevik Russia into a biofueled workers' state. No Marxist historian, Malle cannot fail to acknowledge the achievement. She notes that the Bolsheviks managed to 'support a well equipped army amidst general distress and disorganization', which, given the energetic dimensions of the two past centuries of warfare, must count as rather exceptional: having the Whites and the allied empires arrayed against them – zero fossil fuels versus all the reserves in the world – the Red Army won the war. In this isolated respect, the period from late 1918 to late 1920 was the finest hour of the Soviet state. Trotsky riding on a wood-fuelled train, FDR surfing on an ocean of oil: take your pick.

But there runs, of course, a gulf between forcible deprivation and active renunciation of fossil fuels. As soon as war communism was over, the chimneys smoked again. For as long as it lasted, the absence of coal and oil was experienced not as a blessing but as a curse, more particularly by the men and women forced to collect all that timber. 'Our fuel requirements cannot be satisfied, even partially, without a mass application on a scale hitherto unknown, of labour-power to work on wood', asserted Trotsky. With no access to coal or oil, starved of modern machinery, the state fell back on manual labour. It dared not trust it to step forth voluntarily. Red Army brigades unneeded at the front were sent into forests; working under military command, they were reasonably efficient in cutting the wood. Trotsky saw in them a model for how the economy should be organised under the burning emergency. A rural district should have 'an obligation to supply, for example, in such a time such a number of cubic *sazhens* of wood'; workers and peasants should be sent out for 'repairing railway lines, cutting timber, chopping and bringing up wood to the towns'. Herein lay a seed of the militarisation of labour, perhaps the most scandalous aspect of war communism. It entailed the identification and conscription of suitable workforces, the imposition of labour duties, the exercise of severe martial discipline – a regime that did exist but, according to Malle, 'was not extensively implemented'.

An energetic interpretation of war communism nonetheless suggests itself. Denied fossil fuels, the state resorted to some measure of forced labour – bodies and trees substituting for the concentrated energy from under the ground – in the battle against armies with all the supplies of the earth at their disposal, thereby setting in motion a train to despotism. Might something similar happen again? If we let the imagination run freely along this track, we could conjure up an image of a revolution against fossil capital in one country, which is immediately set upon by hostile neighbours, prompting the state to draft manual labour ... but such a repetition must be considered exceedingly implausible. The renewable energy of today, for a start, has none of the intrinsic deficiency of wood relative to coal and oil. It doesn't take an army of calloused *muzhiks* to collect the sun or the wind. Moreover, one must be Trotskyist enough to presume that no break with fossil capital will ever happen in a single nation; slightly more conceivable is an international scramble, even if not as synchronous as the reaction to Covid-19, when the climate crisis reaches some global breaking point.

For all its unpleasantness, then, war communism would seem to meet the three criteria Kester and Sovacool correctly contend that the US of World War II left unfulfilled. There are additional reasons to think in its terms. It is politics in the key of an emergency turned chronic. 'An habitual, normal regime – an habitual, normal method of work – will not save us now.' This might sound a bit like Boris Johnson or Angela Merkel in the spring of 2020, or the summary for policymakers in any IPCC report, but it is, of course, Trotsky in the *annus horribilis* 1920. It is close to the antithesis of utopia. The early years of the Bolshevik regime 'are no more images of utopia than are rescue teams at mining disasters, which require chains of command and forms of discipline we would find objectionable if we woke up to them every day', Terry Eagleton remarks. For all his brilliance, Eagleton belongs to that species of Marxists who can put out one book per year and still never notice that something is going on in the climate and non-human world, but his simile lends itself to ecological use: global heating is the mining disaster *ne plus ultra.* It calls for rescue operations, not to be idealised as the perfect society. Saving people from asphyxiating in a coal shaft is not to open the door to *Schlaraffenland.* Likewise, 'peace communism' would be a misnomer for the foreseeable future, as it connotes an enjoyable gambol through the meadows.

Now we have argued that a transition away from fossil fuels is in no need of repression à la lockdown but rather conducive to the betterment of peoples' lives. So it is, but there is another side to the coin: a degree of forsaking. In an ideal world, with no material constraints, everyone could hunt whatever they wanted and fly as much as they liked. Perhaps they will one day be able to, after a transitional period that reforests enough of the earth and reinvents aviation on another technical footing (possibly the concentrated fuel captured from the air). But *during the transitional period* there is no escaping outlawing wildlife consumption and terminating mass aviation and phasing out meat and other things considered parts of the good life, and those elements of the climate movement and the left that pretend that none of this needs to happen, that there will be no sacrifices or discomforts for ordinary people, are not being honest. They are being less honest the longer the transition waits. In a programmatic essay, the Salvage Collective has written an elegant obituary for the classical Marxist dream of a humanity liberated in a land of abundance, the last dreamers of which are the comrades fantasising about 'luxury communism': as damaged as this planet already is, nothing suggests that plenitude will be there for the taking. Shortages and trade-offs and aftershocks from the planetary disruption will be on the estate when the anti-capitalist forces finally break through. 'The earth the wretched would – will – inherit, will be in need of an assiduous programme of restoration', the two main forms of which must be rewilding and draw-down. Keeping with the brand, the Collective calls this 'Salvage Communism'. It is a more positively charged term. The content is rather the same: 'all politics must become disaster politics', with the Collective; 'our position is in the highest degree tragic', with Trotsky anno 1920. The ecological crisis is nothing if not tragic.

War communism as practised in Russia between 1918 and 1921 was an attempt to crawl out of the pit, and as Lars T. Lih has showed in a series of meticulous investigations, this required compromises. Turning inside out the picture of ecstatic Bolsheviks leaping into the promised land in those years, he details how they had to swallow one bitter pill after another: recruiting old Tsarist officers to the Red Army; increasing productivity by doling out bonuses and encouraging wage differentiation; accepting the authority of 'bourgeois specialists' – engineers, technicians, managers – in the workplaces; giving up on the ideals of communes and much else. Paradoxically, the emergency dictated some contaminating continuity with the *ancien régime*. There can be no clean break; the new regime 'is bound to draw from the old institutions all that was vital and valuable in them, and harness it on to the new work'. Our bourgeois specialists will be recruited from the oil companies and the start-ups. But here we also reach the final value of war communism as metaphor and analogue: the very negative lessons.

The journey from war communism to tyranny was short to non-existent. The convenient conclusion from this history is that state-led emergency action is always bound to derail into totalitarianism and should therefore be *a priori* excluded. Kester and Sovacool worry that democracy itself will have to be sacrificed during a climate wartime mobilisation: 'the cure may be worse than the disease', a phrase also heard in protests against the lockdown. There comes a point when that position is guilty of underestimating the disease. At that precise point, one must be prepared to *stay with the dilemma*, to adapt a phrase from Donna Haraway: the dilemma of how to execute control measures in an emergency without trampling on democratic rights, but rather by securing, building on and drawing force from them. Neither anarchists nor social democrats recognise this dilemma, but on the phylogenetic tree of socialism, there is one branch that has spent its lifetime mulling over it and never letting go of it as a matter of principle: anti-Stalinist Leninism. Has it produced any straightforward resolution? Of course not. It has only learnt some hard lessons about how wrong things can go when the guard against bureaucratic usurpation is let down. It has, at least in its heterodox families, fretted over the mistakes of the founders – Lenin and Trotsky glorifying hard central power and succumbing to a fortress mentality – and vowed not to repeat them. It has experimented with an 'oddly libertarian Leninism' (Bensaïd) that installs checks on such power and instils vigilance against bureaucracy as a supreme value. For one century, it has ruminated over the question of just when the locomotive of October went off track, what in its internal construction contributed to the wreck and how, or if, it could have been steered more productively, without coming up with an exact manual for how to master the dilemma the next time – because it's in the nature of the dilemma that there can be none. There can only be a set of inviolable principles, first among them to never ever infringe on the freedom of expression and assembly.

No such knowledge bank came out of the US experience of World War II. In the chronic emergency, *this* is the tradition to return to, for if there would ever be a state that took control of trade flows, chased down wildlife traffickers, nationalised fossil fuel companies, organised direct air capture, planned for the economy to cut nearly 10 per cent of emissions per year and did all the other necessary things, we would be on our way out of the emergency. But the journey would obviously be fraught with danger. A state thus expanded could – even if it shrank in other departments: surveillance, military apparatuses, migration control – become bloated. As any transition is hard to conceive without popular-democratic ferment, tensions along a vertical axis may ensue. All this is, of course, purely in the realm of speculation, but we now face the more imminent problem of authoritarian degeneration in periods of symptomatic treatment such as lockdown. Some bourgeois states – those on the far right – will have no compunctions about extending their repressive powers towards proto-fascism. This threat will be dealt with quite extensively in another study.

The future, then, is ecological war communism, in a figurative sense, this being 'only an analogy – but an analogy very rich in content'. It means learning to live without fossil fuels in no time, breaking the resistance of dominant classes, transforming the economy for the duration, refusing to give up even if all the worst-case scenarios come true, rising out of the ruins with the force and the compromises required, organising the transitional period of restoration, staying with the dilemma. It does not mean cosplay re-enactments of the Russian Civil War. That war deposited a poison of brutalised power in the heart of the workers' state, to which it eventually fell victim. Another legacy of the period, however, fared better.

Lenin's passion for wild nature is well known to his biographers (as is Luxemburg's, in a slightly different register; both will be dealt with in some depth elsewhere).

It was turned into practical policy with no delay. Two days after seizing power, the Bolsheviks published their decree 'On Land', which made all forests, waters and minerals the property of the state; half a year later, they promulgated a basic law 'On Forests', which divided the forests of Russia into one exploitable and one protected sector, the latter set aside for what was called 'the preservation of monuments of nature' – monuments, that is, not built by people but by nature itself. At the height of the war in 1919, Bolshevik activists from the besieged town of Astrakhan managed to reach Moscow and appeal to Lenin to approve their plans for a nature reserve in the Volga delta. So he did, explaining that 'conservation of nature is of importance to the entire Republic; I attach urgent significance to it. Let it be declared a national necessity and appreciated by the scale of nationwide importance.' That reserve became the first *zapodevnik* instituted by the Bolsheviks. It was followed by legislation in 1921 that ordered 'significant areas of nature' across the Russian continent to be protected, the idea being not to turn these areas into pleasure grounds, but to set them aside as untrammelled natural systems for the sake of science and nature itself.

The end of the war saw communist biologists fanning out across the Russian landmass to establish new *zapodevniks*. Ecological sciences entered a period of extraordinary effervescence, bringing to the fore such figures as the entomologist V. P. Semenov-Tian-Shanskii, who warned that 'big capital' had caused a mutation in the 'predator nature' of the human species now threatening the continued survival of many animals. In 1922, the hydrologist V.E. Timonov painted a gloomy picture of the effects of exploiting nature for 'the most immediate profits', lamenting:

> The climate is being ruined. The conditions for life are deteriorating. Capping his 'victory' over Nature, man [sic] places amidst the most attractive scenery disgusting billboards, especially perverse forms in this age of 'steam and electricity', the two forces impelling factories to spew their foul-smelling gases into the atmosphere, greatly impairing the enjoyment of nature.

There was a moment when such voices had the ears of the state. It was tragically brief. Stalinism was also an ecological counter-revolution, but it didn't succeed in obliterating the legacy of October. The foremost historian of conservation in the Soviet Union, Douglas Weiner, observes that

> Russians were first to propose setting aside protected territories for the study of ecological communities, and the Soviet government was first to implement that idea. Furthermore, Russians pioneered the suggestion that regional land use could be planned and *degraded landscapes rehabilitated* on the basis of those ecological studies

– ideas that lived on, including in the practice today referred to as rewilding.

More remarkably still, the system of *zapodevniks* survived the Stalin era and the fall of the Soviet Union, so that, one hundred years after October, Russia cordoned off the most land with the highest level of protection – defined as 'strict nature reserves' with no visitors allowed – within any nation on earth. Speak of an enduring gain. Under the headline 'Lenin's Eco-Warriors', the *New York Times* ran an opinion piece expressing some admiration in August 2017:

> For now, at least, Lenin's legacy is preserved and Russia remains the world leader, ahead of Brazil and Australia, in protecting the most land at the highest level. Russian naturalists continue to advance their not-yet-hopeless cause of keeping free a few vast landscapes on this planet where humans do not tread.

Another historical task for ecological Leninism in the twenty-first century: expanding some truly vast landscapes on this planet where humans do not tread.

Ending a boundless imperialism

'There is a universal feeling, a universal fear, that our progress in controlling nature may increasingly help to weave the very calamity it is supposed to protect us from.'

Theodor Adorno, the greatest thinker of the twentieth century, would tremble from a surfeit of *déjà vu* if he were alive. So would his collaborator Max Horkheimer, who put it this way:

> Nature is today more than ever conceived as a mere tool of man [sic]. It is the object of total exploitation that has no aim set by reason, and therefore no limit. Man's boundless imperialism is never satisfied. The dominion of the human race over the earth has no parallel in those epochs of natural history in which other animal species represented the highest forms of organic development. Their appetites were limited by the necessities of their physical existence,

but human appetites have no such limits, for reasons, Horkheimer hastened to add, developed over the course of history.

> Man's avidity to extend his power in two infinities, the microcosm and the universe, does not arise directly from his own nature, but from the structure of society. Just as attacks of imperialistic nations on the rest of the world must be explained on the basis of their internal struggles rather than in terms of their so-called national character, so the totalitarian attack of the human race on anything that it excludes from itself derives from interhuman relationships rather than from innate human qualities.

The domination of nature goes deeper in history than capital. But it does not go deeper than class society. It would take another study to even begin to demonstrate this, as well as to disentangle the analytical content of 'the domination of nature', arguably the key concept of the Frankfurt School; both these tasks will be attempted elsewhere. Here we shall merely ask: is Covid-19 'the revolt of nature', to use Horkheimer's term?

Some liked to see it that way. In March and April 2020, there evolved a genre of pandemic romanticism, as people posted pictures of wild animals asserting themselves where humans had retreated: wild boars descending from the mountains to the streets of Barcelona, lions on the asphalt in South Africa, deer in Japanese subway stations, pumas in Santiago de Chile – #animalstakingover. Sublime as this might seem, zoonotic spillover is, unfortunately, not a viable insurrectionary tactic from persecuted fauna, because it can easily spill back and infect wild animals. Particularly nervous about primates, scientists urged the suspension of ecotourism and field research in or near habitats of great apes when the pandemic started. The apes could lose their breath like any humans. Infectious diseases have already annihilated animal populations – 5,000 gorillas in one Congolese district lost to Ebola in 2003; a fungal pathogen driving amphibians across the world to extinction – and so global sickening cannot be conceived as the revenge or revolt of nature, other than possibly as a blind martyrdom operation: a wild Samson in shackles bringing down the pillars of the roof over the banquet. But any such scene is, of course, in the eyes of the human beholder.

Latourians, posthumanists, new materialists and other hybridists can be counted on to anoint corona with agency. We can look forward to academic articles with titles such as 'Pathogen Performativity: Towards an Understanding of How Viruses Come to Matter' or 'We Have Never Been Healthy: On the Dissolution of the Boundary between Peoples and Parasites' or 'The Agentic Assemblage of Covid-19: Distributed Capacities, Vibrant Monstrosities'. It will be as obfuscating as ever. This episode in the ecological crisis has once again underlined the indelible ontological distinction between humans and non-humans: bats didn't one day tire of their forests. Pangolins didn't offer themselves for sale. The organism known as SARS-CoV-2 has never devised a plan for infiltrating airplanes or borne any onus against anyone. The only agents with intentions in this affair are humans, who can think thoughts like 'if I breed those rats I can sell more of them' or 'there's oil under that swamp'. Humans and humans alone poke a stick in the hornet's nest. What something like the corona crisis could be, however, is the moment when 'human beings become conscious of their own naturalness and call a halt to their own domination of nature', with Adorno, *conscious* here being a keyword. More precisely, zoonotic spillover of this earth-shattering magnitude should make it clear that *defending wild nature against parasitic capital is now human self-defence.* But the conscious organisation of such defence is solely up to humans.

On a bigger scale, one must contemplate, again with Adorno,

> whether humanity is capable of preventing catastrophe. The forms of humanity's own global societal constitutions threaten its life, if a self-conscious global subject does not develop and intervene. The possibility of progress, of averting the most extreme, total disaster, has migrated to this global subject alone. Everything else involving progress must crystallize around it.

Where is that global subject? Who is it? Merely asking such questions is to weigh up the void in which we fumble. So is bringing in Lenin or speaking of war communism: they would never be needed if it weren't for them serving as indexes of the gravity of our ordeal. Precisely in their remoteness from any currently discernible trajectory, not to say their farfetchedness, lies their truth content. An age this bad can only be reflected in extreme contrasting images. Likewise, every concrete measure proposed here and by many others for coming to terms with nature and ending the 'boundless imperialism' against it may well be brushed aside as utopian. They are exactly as utopian as survival.

Neukölln, Berlin, 28 April 2020

Acknowledgements

Massive thanks to Critical Theory in Berlin, that amazing hub of intellectual vibrancy housed by the Humanities and Social Change Center at Humboldt University, whose very generous support made the writing of this book possible. Special thanks to Rahel Jaeggi. As this book is not based on primary research, I must express my profoundest gratitude to all the researchers whose work I have borrowed from and my hope that I have not done violence to it. Any errors are my responsibility solely (as are failures to spot relevant research and data available at the time of writing). Thanks to Sebastian Budgen, who asked me to do this. To Ståle Holgersen, for inspirational conversations. To Wim Carton and Richard Seymour, who read a first draft and gave invaluable comments. To Farahnaz Banyasad, for very special help. Thanks, most of all, to the three sources of joy in my life: Shora Esmailian – my gratitude to whom is not to be expressed in words – and Latifa and Nadim Esmailian Malm.

Notes

1. Corona and Climate

p. 1 **when a virus jumped out of a food market . . .** This summary of the aetiology of Covid-19 is based on the available literature in peer-reviewed journals at the time of writing: Xintian Xu, Ping Chen, Jingfang Wang et al., 'Evolution of the Novel Coronavirus From the Ongoing Wuhan Outbreak and Modeling of Its Spike Protein for Risk of Human Transmission', *Science China* (2020), online first; Fan Wu, Su Zhao, Bin Yu et al., 'A New Coronavirus Associated With Human Respiratory Disease in China', *Nature* 579 (2020): 265–9; Peng Zhou, Xing-Lou Yang, Xian-Guang Wang, 'A Pneumonia Outbreak Associated With a New Coronavirus of Probable Bat Origin', *Nature* 579 (2020): 270–3; Kristian G. Andersen, Andrew Rambaut, W. Ian Lipkin et al., 'The Proximal Origin of SARS-CoV-2', *Nature Medicine*, nature.com, 17 March 2020; Chaolin Huang, Yeming Wang, Xingwang Li et al., 'Clinical Features of Patients Infected With 2019 Novel Coronavirus in Wuhan, China', *The Lancet* 395 (2020): 497–506; Xingguan Li, Junjie Zai, Qiang Zhao, 'Evolutionary History, Potential Intermediate Animal Host, and Cross-Species Analysis of SARS-CoV-2', *Journal of Medical Virology* (2020, online first; Alfonso J. Rodriguez-Morales, D. Katterine Bonilla-Aldana, Graciela Josefina Balbin-Ramon et al., 'History Is Repeating Itself: Probable Zoonotic Spillover as the Cause of the 2019 Novel Coronavirus Epidemic', *Le Infezioni in Medicina* no. 1 (2020): 3–5. See also John Vidal, 'Destroyed Habitat Creates Perfect Condition for Coronavirus to Emerge', *Scientific American*, scientificamerican.com, 18 March 2020. Uncertainties remained about many steps in the aetiology, including the exact role of the Wuhan market: see e.g. Jon Cohen, 'Mining Coronavirus Genomes for Clues to the Outbreak's Origins', *Science*, sciencemag.org, 31 January 2020; David Cyranoski, 'Mystery Deepens Over Animal Source of Coronavirus', *Nature* 579 (2020): 18–19.

p. 2   **And while this was happening, swarms of locusts . . .** Antoaneta Roussi, 'The Battle to Contain Gigantic Locust Swarms', *Nature* 579 (2020): 330; Kaamil Ahmed, 'Locust Crisis Poses a Danger to Millions, Forecasters Warn', *Guardian*, theguardian.com 20 March 2020; Bob Berwyn, 'Locust Swarms, Some 3 Times the Size of New York City, Are Eating Their Way Across Two Continents', *Inside Climate News*, insideclimatenews.org, 22 March 2020.

p. 5   **'Starbucks is not essential' . . .** Anyia Johnson quoted in Emma Ockerman, 'Starbucks Employees Are Begging the Company to Shut Down Stores Because of Coronavirus', *Vice*, vice.com, 19 March 2020.

p. 6   **The US president Donald Trump was initially not fond . . .** Trump quoted in Laurence Darmiento, 'We're at War With COVID-19. What Lessons Can We Learn From World War II?', *Los Angeles Times*, 27 March 2020.

p. 6   **GM and Ford began clearing out idle plants . . .** Neal E. Boudette and Andrew Jacobs, 'Inside G.M.'s Race to Build Ventilators, Before Trump's Attack', *New York Times*, 30 March 2020.

p. 7   **in Sweden, flight attendants from . . .** Niklas Weimer and Vanni Jung Ståhle, 'Från flyget till vården – här utbildas SAS-personalen: "Rena halleljuahistorien" ', *Dagens Nyheter*, dn.se, 1 April 2020.

p. 7   **mass flying relegated to the era BC (Before Coronavirus) . . .** To the knowledge of this author, the new calendar was first proposed in Jeremy Cliffe, 'The Rise of the Bio-surveillance State', *New Statesman*, newstatesman.com, 25 March 2020.

p. 8   **'Not only is this the largest economic shock . . .'** Jeffery Currie, with the post 'global head of commodities' at the firm, quoted in Jillian Ambrose, 'Oil Rig Closures Rising as Prices Hit 18-Year Lows', *Guardian*, 30 March 2020.

p. 8   **During February 2020, the combustion of coal . . .** Lauri Myllyvirta, 'Analysis: Coronavirus Has Temporarily Reduced China's CO_2 Emissions by a Quarter', *Carbon Brief*, carbonbrief.org, 30 March 2020. For updates on emissions trends, *Carbon Brief* is the outstanding source.

p. 10   **The most cited paper on how . . .** Mark A. Delucchi and Mark Z. Jacobson, 'Providing All Global Energy With Wind, Water, and Solar Power, Part II: Reliability, System and Transmission Costs, and Policies', *Energy Policy* 39 (2011), 1178.

p. 10   In the most detailed comparison . . . Laurence Delina, *Strategies for Rapid Climate Mitigation: Wartime Mobilisation as a Model for Action* (Abingdon: Routledge, 2016). This book elaborated the argument outlined in Laurence L. Delina and Mark Diesendorf, 'Is Wartime Mobilisation a Suitable Policy Model for Rapid National Climate Mitigation?', *Energy Policy* 58 (2013): 371–80.

p. 10   In 2011, the year when that paper . . . Lester Brown, Sally G. Bingham, Brent Blackwelder et al., 'Open Letter to President Barack Obama and President Hu Jintao', *350.org*, 22 January 2011.

p. 11   One reader of that book, Bill McKibben . . . Bill McKibben, 'A World at War', *New Republic*, newrepublic.com, 15 August 2016; Peter Weber, 'Bernie Sanders Frames Climate Change as an Urgent Existential War', *The Week*, theweek.com, 14 April 2016; Joe Romm, 'Democratic Platform Calls for WWII-Scale Mobilization to Solve Climate Crisis', *Think Progress*, thinkprogress.org, 22 July 2016; Jeff McMahon, 'Hillary Clinton Plans to Have a "Climate Map Room" in the White House, Podesta Says', *Forbes*, forbes.com, 8 May 2016.

p. 11   One offshoot of the US . . . The Climate Mobilization, *The Victory Plan*, theclimatemobilization.org, n.d.

p. 12   up the ladder to establishment figures like . . . Joseph Stiglitz, 'The Climate Crisis Is Our Third World War. It Needs a Bold Response', *Guardian*, 4 June 2019; Kate Proctor, 'Ed Miliband Calls for "Wartime" Mobilisation to Tackle Climate Crisis', *Guardian*, 23 September 2019.

p. 13   Nor can the state of the science . . . As argued by Erik Schliesser and Eric Winsberg, 'Climate and Coronavirus: The Science Is Not the Same', *New Statesman*, 23 March 2020. Cf. e.g. Nick Chater, 'Facing Up to the Uncertainties of COVID-19', *Nature Human Behaviour*, nature.com, 27 March 2020; Dyani Lewis, 'Is the Coronavirus Airborne? Expert Can't Agree', *Nature*, 2 April 2020.

p. 14   Someone suggested that . . . Barbara Buchner of the Climate Policy Initiative quoted in David Vetter, 'How Coronavirus Could Help Us Fight Climate Change: Lessons From the Pandemic', *Forbes*, 30 March 2020.

p. 14   Some argued that the pandemic . . . Quotations from Eric Galbraith and Ross Otto, 'Coronavirus Response Proves the World Can Act on Climate Change', *The Conversation*, theconversation.com, 23 March 2020. Similar arguments were put forth in David Comerford, 'Here's Why We've Responded to Coronavirus So Wildly Differently to Climate Change', *The Conversation*, 14 March 2020; Nives Dolsak and Aseem Prakash, 'Here's Why Coronavirus and Climate Change Are Different Sorts of Policy Problems', *Forbes*, 15 March 2020.

p. 14 'far-off probability', a 'distant, non-certain threat' ... Eve Andrews, 'Is Life During Coronavirus How We Will Live During Climate Change?', *Grist*, grist.org, 26 March 2020.

p. 14 'is not a problem for future generations ... Galbraith and Otto, 'Coronavirus'.

p. 14 By March 2020, the WHO ... See e.g. World Health Organization, 'Climate Change and Health', who.int, 18 June 2008; *Scientific American*, 'The Impact of Global Warming on Human Fatality Rates', 17 June 2019.

p. 14 The World Meteorological Organization (WMO) estimated ... World Meteorological Organization, *WMO Statement on the State of the Global Climate in 2019*, wmo.int, 29–30.

p. 15 Equally faulty is the assertion ... Made in Comerford, 'Here's Why'.

p. 15 'no single, clear "enemy" ... Andrews, 'Is Life'.

p. 15 'the future is going to be bad ... Comerford, 'Here's Why'.

p. 16 This counted as among the more popular ... E.g. Peter C. Baker, ' "We Can't Go Back to Normal": How Will Coronavirus Change the World?', *Guardian*, 31 March 2020.

p. 17 in scenes of 'apocalyptic devastation' ... Words from Luca Zaia, governor of the Veneto region, in Jon Henley and Angela Giuffrida, 'Two People Die as Venice Floods at Highest Level in 50 Years', *Guardian*, 14 November 2019.

p. 17 600 billion tons of ice crashed ... Isabella Velicogna, Yara Mohajerani, Felix Landerer et al., 'Continuity of Ice Sheet Mass Loss in Greenland and Antarctica From the GRACE and GRACE Follow-on Missions', *Geophysical Research Letters* (2020), online first; Oliver Milman, 'Greenland's Melting Ice Raised Global Sea Level by 2.2 mm in Two Months', *Guardian*, 19 March 2020; Damian Carrington, 'Polar Ice Caps Melting Six Times Faster Than in the 1990s', *Guardian*, 11 March 2020; Gregory S. Cooper, Simon Willock and John A. Dearing, 'Regime Shifts Occur Disproportionately Faster in Larger Ecosystems', *Nature Communications* 11 (2020), 7.

p. 18 Here one observer pointed out ... Jonathan Watts, 'Delay Is Deadly: What Covid-19 Tells Us About Tackling the Climate Crisis', *Guardian*, 24 March 2020.

p. 19 In 2003, Europe boiled ... See e.g. S. Vandentorren, P. Bretin, A. Zeghnoun et al., 'August 2003 Heat Wave in France: Risk Factors for Death of Elderly People Living at Home', *European Journal of Public Health* 16 (2006): 583–91; Daniel Mitchell, Clare Heaviside, Sotiris Vardoulakis et al., 'Attributing Human Mortality During Extreme Heat Waves to Anthropogenic Climate Change', *Environmental Research Letters* 11 (2016): 1–8.

p. 19   **Two heatwaves in the summer of 2019** . . . WMO, *WMO Statement*, 28.

p. 19   **Smoke from the bushfires entered** . . . Nicholas Borchers Arriagada, Andrew J. Palmer, David M.J.S. Bowman et al., 'Unprecedented Smoke-Related Health Burden Associated With the 2019–20 Bushfires in Eastern Australia', *The Medical Journal of Australia* (2020), online first.

p. 19   **in the calculation of one Stanford researcher** . . . Marshall Burke, 'COVID-19 Reduces Economic Activity, Which Reduces Pollution, Which Saves Lives', *Global Food, Environment and Economic Dynamics*, g-feed.com, 8 March 2020.

p. 20   **Bespeaking a degree of absurdity** . . . Jeff McMahon, 'Coronavirus Lockdown May Save More Lives by Preventing Pollution Than by Preventing Infection', *Forbes*, 11 March 2020, quoting François Gemenne.

p. 20   **One estimate in late March suggested** . . . Patrick G. T. Walker, Charles Whittaker, Oliver Watson et al., 'The Global Impact of COVID-19 and Strategies for Mitigation and Suppression', Imperial College London, imperial.ac.uk. 26 March 2020.

p. 22   **The ten countries with the most deaths from Covid-19** . . . Molly Blackall, Damien Gayle, Mattha Bussby et al., 'Covid-19 as It Happened', *Guardian*, 29 March 2020.

p. 22   **As it happens, all of these countries** . . . Hannah Ritchie and Max Roser, 'CO$_2$ and Greenhouse Gas Emissions', *Our World in Data*, ourworldindata.org, December 2019. Note that the EU countries are treated as one territorial unit in these statistics, which does not alter the basic pattern.

p. 22   **One German liberal put it well** . . . Ralf Fücks, 'Weshalb wir aus der Corona-Not keine ökologische Tugend machen sollten', *Die Welt*, welt.de, 16 March 2020.

p. 23   **When the parameters of the climate crisis** . . . For an extensive analysis of this period and the organised sabotage of climate mitigation, see Andreas Malm and The Zetkin Collective, *White Skin, Black Fuel: On the Dangers of Fossil Fascism* (London: Verso, forthcoming).

p. 24   **So suddenly did it strike** . . . Similar contrasts were made by Beth Gardiner, 'Coronavirus Holds Key Lessons on How to Fight Climate Change', *Yale Environment 360*, e360.yale.edu, 23 March 2020; Dolsak and Prakash, 'Here's Why'; Galbraith and Otto, 'Coronavirus'. I owe the arguments in this and the following paragraph to discussions with Ståle Holgersen.

p. 24   **Covid-19 came as an instantaneous and total saturation** . . . Cf. Emiliano Terán Mantovani, 'Coronavirus beyond Coronavirus: Thresholds, Biopolitics and Emergencies', *Undisciplined Environments*, undisciplined-environments.org, 30 March 2020.

p. 25 Suppression of Covid-19 fits into the overarching . . . A similar argument was put forth by Annelise Depoux and François Gemenne, 'De la crise du coronavirus, on peut tirer des leçons pour lutter contre le changement climatique', *Le Monde*, lemonde.fr, 18 March 2020.

p. 26 Left untreated, both afflictions . . . This was pointed out by Howard Kunreuther and Paul Slovic, 'What the Coronavirus Curve Teaches Us About Climate Change', *Politico*, politico.eu, 26 March 2020; Thomas L. Friedman, 'With the Coronavirus, It's Again Trump vs. Mother Nature', *New York Times*, 31 March 2020.

p. 26 'does not justify policies . . . David R. Henderson and John H. Cochrane, 'Climate Change Isn't the End of the World', *Wall Street Journal*, wsj.com, 30 July 2017.

p. 27 A global poll conducted . . . Christoph Dözlitsch, 'Global Study About COVID-19: Dalia Assesses How the World Ranks Their Governments' Response to the Pandemic', *Dalia*, daliaresearch.com, 30 March 2020.

p. 27 Corona and climate could both be framed . . . Discussions of the problems in these terms could be found in Jason Bordoff, 'Sorry, but the Virus Shows Why There Won't Be Global Action on Climate Change', *Foreign Policy*, foreignpolicy.com, 27 March 2020; Dolsak and Prakash, 'Here's Why'.

p. 29 it could offer improvements . . . Cf. Owen Jones, 'Why Don't We Treat the Climate Crisis With the Same Urgency as Coronavirus?', *Guardian*, 5 March 2020.

p. 29 Hence parts of fossil capital and its friends . . . For this strand of climate denial, see Malm and The Zetkin Collective, *White Skin*.

2. Chronic Emergency

p. 31 'Zoonotic spillover' is less of a household term . . . Based on David Quammen, *Spillover: Animal Infections and the Next Human Pandemic* (New York: W. W. Norton, 2012 [all page numbers refer to the e-book edition, the print edition rapidly going out of print when Covid-19 started]), 5, 13, 17–18, 38–9, 293–4, 302–3, 347–8, 574; Richard Levins, Tamara Awerbuch, Uwe Brinkmann et al., 'The Emergence of New Diseases', *American Scientist* 82 (1994), 53–7.

p. 32 Under normal conditions, coronaviruses . . . Following Quammen, *Spillover*, 15–17, 32, 37–8, 386; Raina K. Plowright, Colin R. Parrish, Hamish McCallum et al., 'Pathways to Zoonotic Spillover', *Nature Reviews Microbiology* 15 (2017): 502–10; Raina K. Plowright, Peggy Eby, Peter J.

Hudson et al., 'Ecological Dynamics of Emerging Bat Virus Spillover', *Proceedings of the Royal Society B* 282 (2014), 3.

p. 33 **'the Western world has virtually . . .** US surgeon general William H. Stewart quoted in Levins et al., 'The Emergence', 52. On this golden age optimism, cf. e.g. A. J. McMichael, 'Environmental and Social Influences on Emerging Infectious Diseases: Past, Present and Future', *Philosophical Transactions of the Royal Society B* 359 (2004), 1049–50.

p. 33 **In his airport bestseller from 2018 . . .** Steven Pinker, *Enlightenment Now: The Case for Reason, Science, Humanism, and Progress* (New York: Penguin, 2018), 64–7, 307. Emphasis in original.

p. 34 **He could, for instance, have opened . . .** Kate E. Jones, Nikkita G. Patel, Marc A. Levy et al., 'Global Trends in Infectious Diseases', *Nature* 451 (2008), 990.

p. 34 **A survey published six years later . . .** Katherine F. Smith, Michael Goldberg, Samantha Rosenthal et al., 'Global Rise in Human Infectious Disease Outbreaks', *Journal of the Royal Society Interface* 11 (2014): 1–6.

p. 34 **Since then, the list of pathogens . . .** E.g. Bryony A. Jones, Martha Betson and Dirk U. Pfeiffer, 'Eco-Social Processes Influencing Infectious Disease Emergence and Spread', *Parasitology* 144 (2017): 26–36; Karl Gruber, 'Predicting Zoonoses', *Nature Ecology and Evolution* 1 (2017): 1–4; Rob Wallace, 'Notes on a Novel Coronavirus', *Monthly Review*, mronline.org, 29 January 2020; Quammen, *Spillover*, 13–14, 17–18. The similarity to hurricanes is noted in Rob Wallace, *Big Farms Make Big Flu: Dispatches on Infectious Disease, Agribusiness, and the Nature of Science* (New York: Monthly Review Press, 2016), 38.

p. 35 **By 2019, the scientific literature referred . . .** Jason R. Rohr, Christopher B. Barrett, David J. Civitello et al., 'Emerging Human Infectious Diseases and the Links to Global Food Production', *Nature Sustainability* 2 (2019), 445.

p. 35 **estimated at between two thirds and three fourths . . .** The higher figure can be found in e.g. Lin-Fa Wang and Danielle E. Anderson, 'Viruses in Bats and Potential Spillover to Animals and Humans', *Current Opinion in Virology* 34 (2019), 79; the lower in e.g. Smith et al., 'Global Rise', 3.

p. 35 **That strange new diseases should . . .** Peter Daszak, Andrew A. Cunningham and Alex D. Hyatt, 'Emerging Infectious Diseases of Wildlife – Threats to Biodiversity and Human Health', *Science* 287 (2000), 446.

p. 35 **The pathogens would not come leaping . . .** Quammen, *Spillover*, 17, 36–8, 386.

p. 35 **'When the trees fall . . .** Ibid., 38.

p. 35 The science is agreed: the secular ... See e.g. Daszak et al., 'Emerging', 443; Andrew A. Cunningham, Peter Daszak and James L. N. Wood, 'One Health, Emerging Infectious Diseases and Wildlife: Two Decades of Progress?', *Philosophical Transactions of the Royal Society B* 372 (2016), 1; Christine K. Johnson, Peta L. Hitchens, Pranav S. Pandit et al., 'Global Shifts in Mammalian Population Trends Reveal Key Predictors of Virus Spillover Risk', *Proceedings of the Royal Society B* 287 (2020), 5; and further references below.

p. 36 This is one of the oldest orders ... Angela D. Luis, David T. S. Hayman, Thomas J. O'Shea et al., 'A Comparison of Bats and Rodents as Reservoirs of Zoonotic Viruses: Are Bats Special?', *Proceedings of the Royal Society B* 280 (2013): 1–9; Cara E. Brook and Andrew P. Dobson, 'Bats as "Special" Reservoirs for Emerging Zoonotic Pathogens', *Trends in Microbiology* 23 (2015): 172–80; Karin Schneeberger and Christian C. Voigt, 'Zoonotic Viruses and the Conservation of Bats' in Christian C. Voigt and Tigga Kingston (eds.), *Bats in the Anthropocene: Conservation of Bats in a Changing World* (Cham: Springer, 2016), 275; Wang and Anderson, 'Viruses', 79; Plowright, 'Ecological', 3, 5.

p. 36 All bats have one hallmark ... Guoije Zhang, Christopher Cowled, Zhengli Shi et al., 'Comparative Analysis of Bat Genomes Provides Insight Into the Evolution of Flight and Immunity', *Science* 339 (2013): 456–60; Thomas J. O'Shea, Paul M. Cryan, Andrew A. Cunningham et al., 'Bat Flight and Zoonotic Viruses', *Emerging Infectious Diseases* 20 (2014): 741–5; Aneta Afelt, Christian Devaux, Jordi Serra-Cobo and Roger Frutos, 'Bats, Bat-Borne Viruses, and Environmental Changes', in Heimo Mikkola (ed.), *Bats* (London: Intech Open, 2018), 120–1; Michael Gross, 'Why We Should Care About Bats', *Current Biology* 29 (2019): R1163–5; Schneeberger and Voigt, 'Zoonotic', 264–5; Brook and Dobson, 'Bats', 176–8; Luis et al., 'A Comparison', 2, 6; Plowright et al., 'Ecological', 6; Wang and Anderson, 'Viruses', 85; Quammen, *Spillover*, 394–5.

p. 37 And bats have a second key trait ... Luis et al., 'A Comparison', 2, 5; Johnson et al., 'Global', 8; Plowright et al., 'Ecological', 3; O'Shea et al., 'Bat', 741, 743; Afelt et al., 'Bats, Bat-Borne', 121–2; Quammen, *Spillover*, 393–4. There are other factors too, such as bats alternating between torpor and flight and thereby further boosting their immune defence systems, and their very ancientness, which has allowed for prolonged co-evolution with pathogens; for these and other bat traits, see references above.

p. 38 The Nipah virus spilled ... Bryony A. Jones, Delia Grace, Richard Kock et al., 'Zoonosis Emergence Linked to Agricultural Intensification and Environmental Change', *PNAS* 21 (2013), 8401; Schneeberger and Voigt, 'Zoonotic', 265–75; Wang and Anderson, 'Viruses', 80–3.

p. 38   sending them into the caves of southern China ... See Jane Qiu, 'How China's "Bat Woman" Hunted Down Viruses From SARS to the New Coronavirus', *Scientific American*, 27 April 2020.

p. 38   The discovery of corona in bats ... Jie Cui, Fang Li and Zheng-Li Shi, 'Origin and Evolution of Pathogenic Coronaviruses', *Nature Reviews Microbiology* 17 (2019): 181–92.

p. 38   One team affiliated with ... Simon J. Anthony, Christine K. Johnson, Denise J. Greig et al., 'Global Patterns in Coronavirus Diversity', *Virus Evolution* 3 (2017): 1–15. Cf. Gruber, 'Predicting', 3.

p. 39   Their genetic information is encoded ... Quammen, *Spillover*, 38–9, 302–3, 347–8; James L. N. Wood, Melissa Leach, Lina Waldman et al., 'A Framework for the Study of Zoonotic Disease Emergence and Its Drivers: Spillover of Bat Pathogens as a Case Study', *Philosophical Transactions of the Royal Society B* 367 (2012), 2; Wang and Anderson, 'Viruses', 85. Coronaviruses are not the only RNA viruses, of course – Nipah and Ebola are some of the others – but their propensity for mutations is higher.

p. 40   The outbreak of Covid-19 in China prompted calls ... Huabin Zhao, 'COVID-19 Drives New Threats to Bats in China', *Science* 367 (2020): 1436.

p. 40   General biodiversity should correlate with ... E.g. Toph Allen, Kris A. Murray, Carlos Zambrana-Torrelio et al., 'Global Hotspots and Correlates of Emerging Zoonotic Diseases', *Nature Communications* 8 (2017), 4–6; Anthony et al., 'Global', 11

p. 40   Biologists have put forth the hypothesis ... E.g. Andy Dobson, Isabella Cattadori, Robert D. Holt et al., 'Sacred Cows and Sympathetic Squirrels: The Importance of Biological Diversity to Human Health', *PLOS Medicine* 3 (2006): 0714–18; Felicia Keesing, Lisa K. Belden, Peter Daszak et al., 'Impacts of Biodiversity on the Emergence and Transmission of Infectious Disease', *Nature* 468 (2010): 647–52; P. T. J. Johnson and D. W. Thieltges, 'Diversity, Decoys and the Dilution Effect: How Ecological Communities Affect Disease Risk', *The Journal of Experimental Biology* 213 (2010): 961–70; Richard S. Ostfelt and Felicia Keesing, 'Effects of Host Diversity on Infectious Disease', *Annual Review of Ecology, Evolution, and Systematics* 43 (2012): 157–82; Felicia Keesing and Truman P. Young, 'Cascading Consequences of the Loss of Large Mammals in an African Savanna', *BioScience* 64 (2014): 487–95; Hamish Ian McCallum, 'Lose Biodiversity, Gain Disease', *PNAS* 112 (2015): 8523–4; David J. Civitello, Jeremy Cohen, Hiba Fatima et al., 'Biodiversity Inhibits Parasites: Broad Evidence for the Dilution Effect', *PNAS* 112 (2015): 8667–71; Michael G. Walsh, Siobhan M. Mor, Hindol Maity and Shah Hossain, 'Forest Loss Shapes the Landscape Suitability of

Kyasanur Forest Disease in the Biodiversity Hotspots of the Western Ghats, India', *International Journal of Epidemiology* 48 (2019): 1804–14; Christine K. Johnson, Peta L. Hitchens, Pranav S. Pandit et al., 'Global Shifts in Mammalian Population Trends Reveal Key Predictors of Virus Spillover Risk', *Proceedings of the Royal Society B* 287 (2020): 1–10. Across-the-board evidence: particularly Civitello et al., 'Biodiversity'. For the continued debate, see Chelsea L. Wood, Kevin D. Lafferty, Giuilo DeLeo et al., 'Does Biodiversity Protect Humans Against Infectious Disease?', *Ecology* 95 (2014): 817–32; Taal Levi, Aimee L. Massey, Robert D. Holt et al., 'Does Biodiversity Protect Humans Against Infectious Disease? Comment', *Ecology* 97 (2016): 536–42; Chelsea L. Wood, Kevin D. Lafferty, Giulio DeLeo et al., 'Does Biodiversity Protect Humans Against Infectious Disease? Reply', *Ecology* 97 (2016): 542–5.

p. 41 **It wouldn't be the first time humans** ... Schneeberger and Voigt, 'Zoonotic', 277–8; Quammen, *Spillover*, 28.

p. 42 **Instead we must examine deforestation** ... Dickson Despommier, Brett R. Ellis and Bruce A. Wilcox, 'The Role of Ecotones in Emerging Infectious Diseases', *EcoHealth* 3 (2007): 281–9; Jonathan D. Mayer and Sarah Paige, 'The Socio-Ecology of Viral Zoonotic Transfer' in Sunit Kumar Singh (ed.), *Viral Infections and Global Change* (Hoboken: Wiley, 2013), 80–1; Kimberley Fornace, Marco Liverani, Jonathan Rushton and Richard Coker, 'Effects of Land-Use Changes and Agricultural Practices on the Emergence and Reemergence of Human Viral Diseases' in ibid., 136–7; Christina L. Faust, Hamish I. McCallum, Laura S. P. Bloomfield et al., 'Pathogen Spillover During Land Conversion', *Ecology Letters* 21 (2018): 471–83; David A. Wilkinson, Jonathan C. Marshall, Nigel P. French and David T. S. Hayman, 'Habitat Fragmentation, Biodiversity Loss and the Risk of Novel Infectious Disease Emergence', *Journal of the Royal Society Interface* 15 (2018): 1–10; Rodrick Wallace, Luis Fernando Chaves, Luke R. Bergmann et al., *Clear-Cutting Disease Control: Capital-Led Deforestation, Public Health Austerity, and Vector-Borne Infection* (Cham: Springer, 2018), 24–5; Benny Borremans, Christina Faust, Kezia R. Manlove et al., 'Cross-Species Pathogen Spillover Across Ecosystem Boundaries: Mechanisms and Theory', *Philosophical Transactions of the Royal Society B* 374 (2019): 1–9; Keesing et al., 'Impacts'.

p. 42 **Fragmentation is now the fate** ... Nick M. Haddad, Lars A. Brudvig, Jean Clobert et al., 'Habitat Fragmentation and Its Lasting Impacts on Earth's Ecosystems', *Science Advances* 1 (2015): 1–9.

p. 42 **One group of ecologists has recently** ... Sarah Zohdy, Tona S. Schwartz and Jamie R. Oaks, 'The Coevolution Effect as Driver of Spillover', *Trends in Parasitology* 35 (2019): 399–408 (quotation from 403).

p. 43   A quarter of the world's bat fauna . . . Tigga Kingston, 'Response of Bat Diversity to Forest Disturbance in Southeast Asia: Insights From Long-Term Research in Malaysia' in R. A. Adams and S. C. Pedersen (eds.), *Bat Evolution, Ecology, and Conservation* (New York: Springer, 2013), 170, 177–80.

p. 43   The result seems to be imposition of chronic stress . . . Anne Seltmann, Gábor Á. Czirják, Alexandre Courtiol et al., 'Habitat Disturbance Results in Chronic Stress and Impaired Health Status in Forest-Dwelling Paleotropical Bats', *Conservation Physiology* 5 (2017): 1–14.

p. 43   'pulses of viral excretion' . . . Raina K. Plowright, Alison J. Peel, Daniel G. Streicker et al., 'Transmission of Within-Host Dynamics Driving Pulses of Zoonotic Viruses in Reservoir-Host Populations', *PLOS Neglected Tropical Diseases* 10 (2016): 1–21.

p. 43   as bats deprived of their old habitats . . . Aneta Afelt, Roger Frutos and Christian Devaux, 'Bats, Coronaviruses, and Deforestation: Toward the Emergence of Novel Infectious Diseases?', *Frontiers in Microbiology* 9 (2018): 1–5; Christoph F. J. Meyer, Matthew J. Struebig and Michael R. Willig, 'Responses of Tropical Bats to Habitat Fragmentation, Logging, and Deforestation' in Voigt and Kingston, *Bats*, 81–2; Plowright et al., 'Ecological', 3–4; Quammen, *Spillover*, 417–18.

p. 44   A break occurred in the 1990s . . . This paragraph sticks closely to Thomas K. Rudel, Ruth DeFries, Gregory P. Asner and William F. Laurance, 'Changing Drivers of Deforestation and New Opportunities for Conservation', *Conservation Biology* 23 (2009): 1396–1405 (quotations from 1398, 1400).

p. 45   No more than four commodities . . . Sabine Henders, U. Martin Persson and Thomas Kastner, 'Trading Forests: Land-Use Change and Carbon Emissions Embodied in Production and Exports of Forest-Risk Commodities', *Environmental Research Letters* 10 (2015): 1–14. Cf. e.g. Patrick Meyfroidt, Kimberley M. Carlson, Matthew E. Fagan et al., 'Multiple Pathways of Commodity Crop Expansion in Tropical Forest Landscapes', *Environmental Research Letters* 9 (2014): 1–13.

p. 45   The historical break was visible from above . . . Kemen G. Austin, Mariano González-Roglich, Danica Schaffer-Smith et al., 'Trends in Size of Tropical Deforestation Events Signal Increasing Dominance of Industrial-Scale Drivers', *Environmental Research Letters* 12 (2017): 1–10.

p. 46   palm oil plantations that creep up . . . Seltmann et al., 'Habitat', 2–3.

p. 46   Malaysia and Indonesia together . . . David L. A. Gaveau, Douglas Sheil, Mohammed A. Salim et al., 'Rapid Conversions and Avoided Deforestation: Examining Four Decades of Industrial Plantation Expansion in Borneo', *Nature Scientific Reports* 6 (2016): 1–13; Henders et al., 'Trading', 1; Austin et al., 'Trends', 5; Oliver Pye, 'Commodifying Sustainability: Development, Nature and Politics in the Palm Oil Industry', *World Development* 121 (2019): 218–28; Oliver Pye, 'Agrarian Marxism and the Proletariat: A Palm Oil Manifesto', *The Journal of Peasant Studies* (2019), online first, 8–10.

p. 46　In Sabah, one researcher . . . Pye, 'Commodifying', 224.

p. 46　'Plantation companies and their shareholders' . . . Gaveau et al., 'Rapid', 8. Governments are often shareholders in these companies, it should be noted.

p. 47　Southeast Asia has seen an epic . . . Derek Byerlee, 'The Fall and Rise Again of Plantations in Tropical Asia: History Repeated?', *Land* 3 (2014): 574–97.

p. 47　Hence the region is the theatre . . . H. E. Field, 'Bats and Emerging Zoonoses: Henipaviruses and SARS', *Zoonoses and Public Health* 56 (2009): 278–84.

p. 47　Deforestation-induced stress has been reported . . . David Costantini, Gábor Á. Czirják, Paco Bustamante et al., 'Impacts of Land Use on an Insectivorous Tropical Bat: The Importance of Mercury, Physio-Immunology and Trophic Position', *Science of the Total Environment* 671 (2019): 1077–85; cf. K. Waiyasusri, S. Yumuang and S. Chotpantarat, 'Monitoring and Predicting Land Use Changes in the Huai Tap Salao Watershed Area, Uthaithani Province, Thailand, Using the CLUE-s Model', *Environmental Earth Sciences* 75 (2016): 1–16.

p. 47　but chiropterologists recognise . . . E.g. Kingston, 'Response', 180; Meyer et al., 'Responses', 64; Field, 'Bats', 282.

p. 47　The Cockpit Country in the heart . . . Field observations by the author in 2017.

p. 48　It attracts a steady stream of protests . . . See e.g. Esther Figueroa, 'Cockpit Country Still Under Threat From Bauxite Mining', *Jamaica Gleaner*, jamaica-gleaner.com, 28 July 2019; Paul Clarke and Andrew Williams, 'Noranda: We All Want to Protect Cockpit Country', *Jamaica Gleaner*, 18 September 2019; *Loop Jamaica*, 'Entertainers Lead Proposed March to Parliament About Cockpit Country', loopjamaica.com, 18 September 2019; and the wonderful short film by Esther Figueroa, the Jamaican film-maker and activist documenting the struggle around the Cockpit Country, '#ClimateStrike_SaveCockpitCountry', YouTube, 3 October 2019.

p. 48　When the Covid-19 pandemic . . . The Jamaica Environment Trust in *Loop Jamaica*, 'Coronavirus: Leave the Bats Alone – JET', 10 April 2020.

p. 49   **This virus, of another family** . . . David W. Redding, Peter M. Atkinson, Andrew A. Cunningham et al., 'Impacts of Environmental and Socio-Economic Factors of Emergence and Epidemic Potential of Ebola in Africa', *Nature Communications* 10 (2019), 2; Derek Gatherer, 'The Unprecedented Scale of the West African Ebola Virus Disease Outbreak Is Due to Environmental and Sociological Factors, Not Special Attributes of the Currently Circulating Strain of the Virus', *BMJ-Evidence Based Medicine* 20 (2015): 28; Robert G. Wallace, Marius Gilbert, Rodrick Wallace et al., 'Did Ebola Emerge in West Africa by a Policy-Driven Phase Change in Agroecology?', in Robert G. Wallace and Rodrick Wallace (eds.), *Neoliberal Ebola: Modeling Disease Emergence From Finance to Forest and Farm* (Cham: Springer, 2016), 1, 4; Wallace, *Big*, 326.

p. 49   **'non-virological' development** . . . Gatherer, 'The Unprecedented'. The factors proposed by this author, including 'funerary practices', appear almost flippant alongside Wallace's rigorous investigations. But cf. also Daniel G. Bausch and Lara Schwarz, 'Outbreak of Ebola Virus Disease in Guinea: Where Ecology Meets Economy', *PLOS Neglected Tropical Diseases* 8 (2014): 1–5.

p. 49   **In the half-decade before the disaster** . . . Wallace et al., 'Did Ebola', 2–9; Wallace, *Big*, 328–30 (World Bank quoted in Wallace *Big*, 328; the other quotations from 333, 327, 330, with emphasis added). Note that the article included in *Big* was originally published as Robert G. Wallace, Richard Kock, Luke Bergmann et al., 'Did Neoliberalizing West African Forests Produce a New Niche for Ebola?', *International Journal of Health Services* 46 (2016): 149–65. Another study independently reached essentially the same conclusion about the role of deforestation: Maria Cristina Rulli, Monia Santini, David T. S. Hayman and Paolo D'Odorico, 'The Nexus Between Forest Fragmentation in Africa and Ebola Virus Disease Outbreaks', *Nature Scientific Reports* 7 (2016): 1–8.

p. 50   **'If landscapes, and by extension** . . . Wallace, *Big*, 331.

p. 50   **It is exported to cosmetic** . . . Pye, 'Commodifying', 221; Henders et al., 'Trading'; Meyfroidt et al., 'Multiple'.

p. 50   **scholars have borrowed the concept** . . . E.g. Yang Yu, Kuishuang Feng and Klaus Hubacek, 'Tele-Connecting Local Consumption to Global Land Use', *Global Environmental Change* 23 (2013): 1178–86. One version is 'tele-coupling': Christina Prell, Laixiang Sun, Kuishuang Feng et al., 'Uncovering the Spatially Distant Feedback Loops of Global Trade: A Network and Input-Output Approach', *Science of the Total Environment* 586 (2017): 401–8.

p. 51 **Remote demand for products** ... Ruth S. DeFries, Thomas Rudel, Maria Uriarte and Matthew Hansen, 'Deforestation Driven by Urban Population Growth and Agricultural Trade in the Twenty-First Century', *Nature Geoscience* 3 (2010): 178–81. Cf. e.g. Eric F. Lambin and Patrick Meyfroidt, 'Global Land Use Change, Economic Globalization, and the Looming Land Scarcity', *PNAS* 108 (2011): 3465–72.

p. 51 **If one calculates the amount of land** ... Yu et al., 'Tele-Connecting'. On Europe as epicentre, see also the telling map in Prell et al., 'Uncovering', 405.

p. 51 **as large as the entire surface area** ... Christian Dorninger and Alf Hornborg, 'Can EEMRIO Analyses Establish the Occurrence of Ecologically Unequal Exchange?', *Ecological Economics* 119 (2015), 417.

p. 52 **a body of literature has demonstrated** ... E.g. Andrew K. Jorgenson, 'Unequal Ecological Exchange and Environmental Degradation: A Theoretical Proposition and Cross-National Study of Deforestation, 1990–2000', *Rural Sociology* 71 (2006): 685–712; John M. Shandra, Christopher Leckband and Bruce London, 'Ecologically Unequal Exchange and Deforestation: A Cross-National Analysis of Forestry Export Flows', *Organization and Environment* 22 (2009): 293–310; Andrew K. Jorgensen, Christopher Dick and Kelly Austin, 'The Vertical Flow of Primary Sector Exports and Deforestation in Less-Developed Countries: A Test of Ecologically Unequal Exchange Theory', *Society and Natural Resources* 23 (2010): 888–97; Andrew K. Jorgenson, 'World-Economic Integration, Supply Depots, and Environmental Degradation: A Study of Ecologically Unequal Exchange, Foreign Investment Dependence, and Deforestation in Less Developed Countries', *Critical Sociology* 36 (2010): 453–77; Kelly Austin, 'The "Hamburger Connection" as Ecologically Unequal Exchange: A Cross-National Investigation of Beef Exports and Deforestation in Less-Developed Countries', *Rural Sociology* 75 (2010): 270–99; Kelly Austin, 'Coffee Exports as Ecological, Social, and Physical Unequal Exchange: A Cross-National Investigation of the Java Trade', *International Journal of Comparative Sociology* 53 (2012): 155–80; Mark D. Noble, 'Chocolate and the Consumption of Forests: A Cross-National Examination of Ecologically Unequal Exchange in Cocoa Exports', *Journal of World-Systems Research* 23 (2017): 236–68; Michael Restivo, John M. Shandra and Jamie M. Sommer, 'Exporting Forest Loss? A Cross-National Analysis of the United States Export-Import Bank Financing in Low- and Middle-Income Nations', *Journal of Environment and Development* 29 (2020): 245–69.

p. 53 **in 2012, one pathbreaking study** ... M. Lenzen, D. Moran, K. Kanemoto et al., 'International Trade Drives Biodiversity Threats in

Developing Nations', *Nature* 486 (2012): 109–12. Cf. e.g. Sabine Henders, Madelene Ostwald, Vilhelm Verendel and Pierre Ibisch, 'Do National Strategies Under the UN Biodiversity and Climate Conventions Address Agricultural Commodity Consumption as Deforestation Driver?', *Land Use Policy* 70 (2018): 580–90.

p. 53   (When these words are written . . . Johns Hopkins University and Medicine, Coronavirus Resource Center, coronavirus.jhu.edu, 13 April 2020.

p. 53   A suite of studies confirming . . . E.g. Franz Essl, Marten Winter and Petr Pysek, 'Trade Threat Could Be Even More Dire', *Nature* 487 (2012): 39; Daniel Moran and Keiichiro Kanemoto, 'Identifying Species Threat Hotspots From Global Supply Chains', *Nature Ecology and Evolution* 1 (2017): 1–5; Francseca Verones, Daniel Moran, Konstantin Stadler et al., 'Resource Footprints and Their Ecosystem Consequences', *Nature Scientific Reports* 7 (2017): 1–11; Thomas Wiedmann and Manfred Lenzen, 'Environmental and Social Footprints of International Trade', *Nature Geoscience* 11 (2018): 314–21; Alexandra Marques, Inês S. Martins, Thomas Kastner et al., 'Increasing Impacts of Land Use on Biodiversity and Carbon Sequestration Driven by Population and Economic Growth', *Nature Ecology and Evolution* 3 (2019): 628–37; Abhishek Chaudhary and Thomas M. Brooks, 'National Consumption and Global Trade Impacts on Biodiversity', *World Development* 121 (2019): 178–87. 60 per cent: Widmann and Lenzen, 'Environmental', 316; closer to one third: Marques et al, 'Increasing'; Chaudhary and Brooks, 'National', 178.

p. 53   One critical nuance has been added . . . Harry C. Wilting, Aafke M. Schipper, Michel Bakkenes et al., 'Quantifying Biodiversity Losses Due to Human Consumption: A Global-Scale Footprint Analysis', *Environmental Science and Technology* 51 (2017): 3298–3306 (quotation from 3304).

p. 54   in early March 2020, *Nature Communications* published a model study . . . Leonardo Suveges Moreira Chaves, Jacob Fry, Arunima Malik et al., 'Global Consumption and International Trade in Deforestation-Associated Commodities Could Influence Malaria Risk', *Nature Communications* 11 (2020): 1–10 (quotation from 5).

p. 55   pronounced version of Wallace's upside-down map . . . Robert G. Wallace, Luke Bergmann, Richard Kock et al., 'The Dawn of Structural One Health: A New Science Tracking Disease Emergence Among the Circuits of Capital', *Social Science and Medicine* 129 (2015), 70; Wallace et al., *Clear-Cutting*, 32. The factor of investment, or capital accumulation as such, has been integrated in at least one statistical study of land appropriation, unsurprisingly showing it to be an even larger phenomenon than if only trade flows are measured: Luke Bergmann and Mollie Holmberg, 'Land in Motion', *Annals of the American Association of Geographers* 106 (2016): 932–56.

p. 55 One of the most comprehensive studies . . . 'mildly significant': De Fries et al., 'Deforestation', 179; 'non-significant': Jorgenson, 'Unequal', 702; positive association: Shandra et al., 'Ecologically'; Jorgenson et al., 'The Vertical'.

p. 55 sometimes counterintuitive . . . DeFries et al., 'Deforestation'; Rudel et al., 'Changing', 1399–1401. On meat consumption as a driver of infectious disease, cf. Rohr et al., 'Emerging', 445.

p. 56 While the towns sprawl . . . Sohel Ahmed, Julio D. Dávila, Adriana Allen et al., 'Does Urbanization Make Emergence of Zoonoses More Likely? Evidence, Myths and Gaps', *Environment and Urbanization* 31 (2019): 443–60; Fornace et al., 'Effects', 142–3.

p. 56 As of 2016, 301 terrestrial mammal . . . William J. Ripple, Katharine Abernethy, Matthew G. Betts et al., 'Bushmeat Hunting and Extinction Risk to the World's Mammals', *Royal Society Open Science* 3 (2016): 1–16; Simon Mickleburgh, Kerry Waylen and Paul Racey, 'Bats as Bushmeat: A Global Review', *Oryx* 43 (2009): 217–34.

p. 56 Rodentia is amply . . . Ripple et al., 'Bushmeat'; Donna-Mareè Cawthorn and Louwrens C. Hoffman, 'Controversial Cuisine: A Global Account of the Demand, Supply and Acceptance of "Unconventional" and "Exotic" Meats', *Meat Science* 120 (2016), 29; Johnson et al., 'Global'.

p. 56 But most of the species thus threatened are primates . . . On primates and other mammals as sources of zoonoses, see Barbara A. Han, Andrew M. Kramer and John M. Drake, 'Global Patterns of Zoonotic Disease in Mammals', *Trends in Parasitology* 32 (2016): 565–77.

p. 57 Indeed, hunters of bushmeat are frequently . . . Nathan D. Wolfe, Peter Daszak, A. Marm Kilpatrick and Donald S. Burke, 'Bushmeat Hunting, Deforestation, and Prediction of Zoonotic Emergence', *Emerging Infectious Diseases* 11 (2005), 1824.

p. 57 Of the infectious diseases spilled . . . Keesing et al., 'Impacts', 651.

p. 57 In the teak forests . . . Tierra Smiley Evans, Theingi Win Myat, Pyaephyo Aung et al., 'Bushmeat Hunting and Trade in Myanmar's Central Teak Forests: Threats to Biodiversity and Human Livelihoods', *Global Ecology and Conservation* (2020), online first.

p. 57 Many hunters capture bushmeat . . . For the complexities of this demography, see e.g. Martin Reinhardt Nielsen, Mariève Pouliot, Henrik Meilby et al., 'Global Patterns and Determinants of the Economic Importance of Bushmeat', *Biological Conservation* 215 (2017): 277–87; Noëlle F. Kümpel, E. J. Millner-Gulland, Guy Cowlishaw and J. Marcus

Rowcliffe, 'Incentives for Hunting: The Role of Bushmeat in the Household Economy in Rural Equatorial Guinea', *Human Ecology* 38 (2010): 251–64; Matthew S. Rogan, Jennifer R. B. Miller, Peter A. Lindsey and J. Weldon McNutt, 'Socioeconomic Drivers of Illegal Bushmeat Hunting in a Southern African Savanna', *Biological Conservation* 226 (2018): 24–31.

p. 58 Deforestation prises open . . . For a fine-grained analysis of how this works, see J. R. Poulsen, C. J. Clark, G. Mavah and P. W. Elkan, 'Bushmeat Supply and Consumption in a Tropical Logging Concession in Northern Congo', *Conservation Biology* 23 (2009): 1597–1608. On the general pattern, see further Wolfe et al., 'Bushmeat'; Diana Bell, Scott Roberton and Paul R. Hunter, 'Animal Origins of the SARS Coronavirus: Possible Links With the International Trade in Small Carnivores', *Philosophical Transactions of the Royal Society B* 359 (2004): 1112; Ripple et al., 'Bushmeat', 6; Fornace et al., 'Effects', 137.

p. 58 The first study of a commodity chain . . . N. Tagg, N. Maddison, J. Dupain et al., 'A Zoo-Led Study of the Great Ape Bushmeat Commodity Chain in Cameroon', *International Zoo Yearbook* 52 (2018): 182–93 (quotation from 182). For a similar analysis of a chain for pangolin meat, see Maxwell Kwame Boakye, Antoinette Kotzé, Desirée Lee Dalton and Raymond Jansen, 'Unravelling the Pangolin Bushmeat Commodity Chain and the Extent of Trade in Ghana', *Human Ecology* 44 (2016): 257–64. On the flow of bushmeat to urban markets in central Africa, see Nathalie van Vliet and Prosper Mbazza, 'Recognizing the Multiple Reasons for Bushmeat Consumption in Urban Areas: A Necessary Step Toward the Sustainable Use of Wildlife for Food in Central Africa', *Human Dimensions of Wildlife* 16 (2011), 47-8.

p. 58 Some bushmeat is upgraded . . . Mickleburgh et al., 'Bats', 225; Meredith A. Barrett and Jonah Ratsimbazafy, 'Luxury Bushmeat Trade Threatens Lemur Conservation', *Nature* 461 (2009): 470.

p. 58 a timely survey in *Science* . . . Brett R. Scheffers, Brunno F. Oliveira, Ieuan Lamb and David P. Edwards, 'Global Wildlife Trade Across the Tree of Life', *Science* 366 (2019): 71–6 (quotations from 71).

p. 59 Standard models in neoclassical economics . . . The paper where this argument was first brilliantly made is Franck Courchamp, Elena Angulo, Philippe Rivalan et al., 'Rarity Value and Species Extinction: The Anthropogenic Allee Effect', *PLOS Biology* 4 (2006): 2405–10 ('extinction vortex', e.g. 2405–6). For further elaborations and empirical testing, see e.g. Elena Angulo, Anne-Laure Deves, Michel Saint Jalmes and Franck Courchamp, 'Fatal Attraction: Rare Species in the Spotlight', *Proceedings of the Royal Society B* 276 (2009): 1331–37; Pierline Tournant, Liana Joseph,

Koichi Goka and Franck Courchamp, 'The Rarity and Overexploitation Paradox: Stag Beetle Collections in Japan', *Biodiversity and Conservation* 21 (2012): 125–40; Jessica A. Lyons and Daniel J. D. Natusch, 'Effects of Consumer Preferences for Rarity on the Harvest of Wild Populations Within a Species', *Ecological Economics* 93 (2013): 278–83; Alex Aisher, 'Scarcity, Alterity and Value Decline of the Pangolin, the World's Most Trafficked Mammal', *Conservation and Society* 14 (2016): 317–29. The general logic is noted in Scheffers et al., 'Global', 74.

p. 60   Here it works the other way around. Suppliers . . . Vanda Felbab-Brown, *The Extinction Market: Wildlife Trafficking and How to Counter It* (New York: Oxford University Press, 2017), 52, 90–3, 98–9, 108, 165.

p. 61   'just a matter of time' . . . Afelt et al., 'Bats, Bat-Borne', 126.

p. 61   In what must be one of the more clairvoyant . . . Hongying Li, Emma Mendelsohn, Chen Zong et al., 'Human-Animal Interactions and Bat Coronavirus Spillover Potential Among Rural Residents in Southern China', *Biosafety and Health* 1 (2019): 84–90. But equally clairvoyant was Yi Fan, Kai Zhao, Zheng-Li Shi and Peng Zhou, 'Bat Coronaviruses in China', *Viruses* 11 (2019): 1–14.

p. 61   prefiguring the sequence of its successor . . . See e.g. Susanna K. P. Lau, Patrick C. Y. Woo, Kenneth S. M. Li et al., 'Severe Acute Respiratory Syndrome Coronavirus-Like Virus in Chinese Horseshoe Bats', *PNAS* 102 (2005): 14040–45; Wendong Li, Zhengli Shi, Meng Yu et al., 'Bats Are Natural Reservoirs of SARS-Like Coronavirus', *Science* 310 (2005): 676–8; Quammen, *Spillover*, 186–9, 209–17.

p. 62   Capitalist development has been as unkind to bats . . . Libiao Zhang, Guangjian Zhu, Gareth Jones and Shuyi Zhang, 'Conservation of Bats in China: Problems and Recommendations', *Oryx* 43 (2009): 179–82; Quan Liu, Lili Cao and Xing-Quan Zhu, 'Major Emerging and Re-emerging Zoonoses in China: A Matter of Global Health and Socioeconomic Development for 1.3 Billion', *International Journal of Infectious Diseases* 25 (2014): 65–72; Tong Wu, Charles Perrings, Ann Kinzig et al., 'Economic Growth, Urbanization, Globalization, and the Risks of Emerging Infectious Diseases in China: A Review', *Ambio* 46 (2017), 19–21.

p. 62   so called because the stalls . . . Paul Jackson, 'Fleshy Traffic, Feverish Borders: Blood, Birds, and Civet Cats in Cities Brimming With Commodities', in S. Harris Ali and Roger Keil (eds.), *Networked Disease: Emerging Infections in the Global City* (Chichester: Blackwell, 2008), 293; Quammen, *Spillover*, 207–9.

p. 62   The richest wanted the most precious commodities . . . Quammen, *Spillover*, 207; Jingjing Yuan, Yonglong Lu, Xianghui Cao and Haotian

Cui, 'Regulating Wildlife Conservation and Food Safety to Prevent Human Exposure to Novel Virus', *Ecosystem Health and Sustainability* 6 (2020), 1. Cf. Mike Davis, *The Monster at Our Door: The Global Threat of Avian Flu* (New York: Henry Holt, 2006), 59–60.

p. 63 One study from the interlude ... Li Zhang and Feng Yin, 'Wildlife Consumption and Conservation Awareness in China: A Long Way to Go', *Biodiversity and Conservation* 23 (2014): 2371–81.

p. 63 the typical customer was a young, rich ... Li Zhang, Ning Hua and Shan Sun, 'Wildlife Trade, Consumption and Conservation Awareness in Southeastern China', *Biodiversity and Conservation* 17 (2008): 1493–1516.

p. 63 Patrons of wet markets testified ... Shuru Zhong, Mike Crang and Guojun Zeng, 'Constructing Freshness: The Vitality of Wet Markets in Urban China', *Agriculture and Human Values* 37 (2020): 175–85 (quotation from 180).

p. 63 And the wet markets were ... E.g. Dennis Normile and W. Li, 'Researchers Tie Deadly Virus SARS to Bats', *Science* 309 (2005), 2155; Paul K. S. Chan and Martin C. W. Chan, 'Tracing the SARS-Coronavirus', *Journal of Thoracic Disease* 5 (2013): S118–S121; Quammen, *Spillover*, 212–20.

p. 64 'eating wildlife as food ... Zhang and Yin, 'Wildlife', 2379.

p. 64 One prominent victim was the pangolin ... This draws on the excellent Alex Aisher, 'Scarcity'; Vincent Nijman, Ming Xia Zhang and Chris R. Shepherd, 'Pangolin Trade in the Mong La Wildlife Market and the Role of Myanmar in the Smuggling of Pangolins Into China', *Global Ecology and Conservation* 5 (2016): 118–26; Neha Thirani Bagri, 'China's Insatiable Thirst for Pangolin Scales Is Fed by an International Black Market', *Quartz*, qz.com, 30 December 2016; Wufei Yu, 'Coronavirus: Revenge of the Pangolins?', *New York Times*, 5 March 2020; Scheffers et al., 'Global', 71, 74.

p. 65 One team claimed to have mapped ... Zhou et al., 'A Pneumonia'; Zhong Sun, Karuppiah Thilakavathy, S. Suresh Kumar et al., 'Potential Factors Influencing Repeated SARS Outbreaks in China', *International Journal of Environmental Research and Public Health* 17 (2020), 4; Kangpeng Xiao, Junqiong Zhai, Yaoyu Feng et al., 'Isolation and Characterization of 2019-nCoV-like Coronavirus From Malayan Pangolins', *BioRxiv*, biorxiv. org, 20 February 2020; note that, as all papers published on this site, this one is a preprint that has not undergone peer review.

p. 65 Other intermediate species have, as of this writing ... This paragraph rests on Sun et al., 'Potential' ('incubation bed': quotation from 7). The Huanong Brothers: Wang Chen, 'Coronavirus Outbreak Reignites Debate

Over Wild Animal Consumption', *Diálogo Chino*, dialogochino.net, 10 February 2020 (for a taste, type 'Huanong Brothers' into YouTube). Rate of RNA evolution: David M. Morens, Peter Daszak and Jeffery K. Taubenberger, 'Escaping Pandora's Box: Another Novel Coronavirus', *The New England Journal of Medicine* 382 (2020), 1293. Cf. also D. Katterine Bonilla-Aldana, Kuldeep Dhama and Alfonso J. Rodriguez-Morales, 'Revisiting the One Health Approach in the Context of COVID-19: A Look Into the Ecology of This Emerging Disease', *Advances in Animal and Veterinary Sciences* 8 (2020): 234–7.

p. 66 **There are now more tigers held captive . . .** Alex Hannaford, 'The Tiger Next Door: America's Backyard Big Cats', *The Observer*, theguardian.com, 10 November 2019.

p. 66 **After China, the US is the second market . . .** Felbab-Brown, *The Extinction*, 8; Center for Biological Diversity, 'Legal Action Initiated to Force Trump Administration to Halt U.S. Pangolin Trade', biologicaldiversity.org, 13 November 2019; Elly Pepper, 'NRDC and Allies Sue Trump Administration to Protect Pangolins', *The Natural Resources Defence Council*, nrdc.org, 22 January 2020.

p. 67 **The rule is very general . . .** Felbab-Brown, *The Extinction*, 9, 38, 53–5, 77, 137.

p. 67 **In the last years BC, one could find . . .** E.g. Tripadvisor reviews of Mongo's Restaurant, Düsseldorf, tripadvisor.com; Karen Thue, 'Narvesen skal selge pølser med krokodillekjøtt – dyrebeskyttelsen reagerer', *Dagbladet*, dagbladet.no, 19 June 2013; Robert Cederholm, 'Försäljningen av exotiskt kött ökar kraftigt i Sverige', *SVT Nyheter*, svt. se, 18 April 2013; Karolina Vikingsson, 'Maten ska smaka pyton', *Göteborgs-Posten*, gp.se, 16 January 2013; Flora Wiström, 'Får det vara en bit pytonorm?', *Flora*, flora.baaam.se, 5 January 2020. Abalone in California: Courchamp, 'Rarity', 2408. For a piquant global survey, see Broglia and Kapel, 'Changing'.

p. 67 **What really set off the vortexes in China . . .** Zhang et al., 'Wildlife'; Felbab-Brown, *The Extinction*, 7–8, 56.

p. 67 **Candace Chen has landed a job . . .** Ling Ma, *Severance* (New York: Farrar, Straus and Giroux, 2018) (quotations from 84, 148, 276). For an excellent reading of the novel, see Jane Hu, ' "Severance" Is the Novel of Our Current Moment – But Not for the Reasons You Think', *The Ringer*, theringer.com, 18 March 2020.

p. 69 **starting with the Black Death . . .** Mark Harrison, *Contagion: How Commerce Has Spread Disease* (New Haven: Yale University Press, 2012), 3–18.

p. 70 **infectious disease still operated under a restraint** . . . Brian D. Gushulak and Douglas W. MacPherson, 'Global Travel, Trade, and the Spread of Viral Infections', in Sing, *Viral*, 116–17.

p. 70 **When steamboats were launched** . . . Harrison, *Contagion*, 81, 107, 119–20, 140–1; Gushulak and McPherson, 'Global', 117; Andrew Cliff and Peter Haggett, 'Time, Travel and Infection', *British Medical Bulletin* 69 (2004), 88–90.

p. 71 **One expert on the fever** . . . William Pym quoted in Harrison, *Contagion*, 87; further 111–12.

p. 71 **In 1832, a cholera pandemic** . . . Harrisson, *Contagion*, 107–15, 139–41, 174–93; Cliff and Hagget, 'Time', 95–6; McMichael, 'Environmental', 1050.

p. 71 **Measles would never have made** . . . Cliff and Hagget, 'Time', 92–4.

p. 72 **Their historical distinctions pertained** . . . Harrison, *Contagion*, e.g. xv, 211.

p. 72 **In 1918, a virus spilled** . . . This paragraph draws on Jeffery K. Taubenberger, Ann H. Reid and Thomas G. Fanning, 'Capturing a Killer Flu Virus', *Scientific American* 292 (2005): 62–71; Davis, *The Monster*, 3, 24–34, 152; Howard Phillips, 'Influenza Pandemic', *International Encyclopedia of the First World War*, encyclopedia.1914-1918-online.net, 8 January 2017; David K. Patterson and Gerald F. Pyle, 'The Diffusion of Influenza in Sub-Saharan Africa During the 1918–1919 Pandemic', *Social Science and Medicine* 17 (1983): 1299–1307; Craig T. Palmer, Lisa Sattenspiel and Chris Cassidy, 'Boats, Trains, and Immunity: The Spread of the Spanish Flu on the Island of Newfoundland', *Newfoundland and Labrador Studies* 22 (2007): 1719–26; Tom Dicke, 'Waiting for the Flu: Cognitive Inertia and the Spanish Influenza Pandemic of 1918–19', *Journal of the History of Medicine and Allied Sciences* 70 (2015): 195–217; Paul Farmer, 'Ebola, the Spanish Flu, and the Memory of Disease', *Critical Inquiry* 46 (2019): 56–70; Amir Afkhami, 'Compromised Constitutions: The Iranian Experience With the 1918 Influenza Pandemic', *Bulletin of the History of Medicine* 77 (2003): 367–92; Wladimir J. Alonso, Francielle C. Nascimento, Rodolfo Acuña-Soto et al., 'The 1918 Influenza Pandemic in Florianopolis: A Subtropical City in Brazil', *Vaccine* 295 (2011): 16–20.

p. 72 **The laconic log book of the *HMS Mantua*** . . . Royal Navy Log Books of the World War I Era, 'HMS Mantua – March 1915 to January 1919, 10th Cruiser Squadron Northern Patrol, British Home Waters, Central Atlantic Convoys', *Naval History*, naval-history.net, 2 November 2014.

p. 73 **'as though the colonial transportation network** . . . Patterson and Pyle, 'The Diffusion', 1302.

p. 73   'it was, in sum, a pandemic . . . Phillips, 'Influenza', 4.

p. 73   Indeed, transcontinental flights are so fast . . . S. Harris Ali, 'Globalized Complexity and the Microbial Traffic of New and Emerging Infectious Disease Threats', in Jennifer Gunn (ed.), *Influenza and Public Health: Learning from Past Pandemics* (London: Earthscan, 2010), 19–29; George J. Armelagos, 'The Viral Superhighway', *Sciences* 38 (1998): 24–9; Gushulak and MacPherson, 'Global', 119, 126; Cliff and Hagget, 'Time', 87–91; Quammen, *Spillover*, 183–6; Davis, *The Monster*, 70–1. Covid-19 is briefly put into this context of post-steam take-off in Morens et al., 'Escaping', 1294.

p. 74   Once it had spread through Wuhan . . . This follows the remarkable reconstruction of travel patterns made by four *New York Times* journalists, based on data tracking millions of cell phones: Jin Wu, Weiyi Cai, Derek Watkins and James Glanz, 'How the Virus Got Out', *New York Times*, 22 March 2020.

p. 74   which explains the peculiar timeline . . . As noted by e.g. Rodrigo Fracalossi de Moraes, 'In Practice, There Are Two Pandemics: One for the Well-Off and One for the Poor', *Global Policy Journal*, globalpolicyjournal.com, 20 March 2020.

p. 75   One could mention ecotourism . . . Michael P. Muehlenbein and Marc Ancrenaz, 'Minimizing Pathogen Transmission at Primate Ecotourism Destinations: The Need for Input From Travel Medicine', *Journal of Travel Medicine* 16 (2009: 229–32; Courchamp et al., 'Rarity', 2408.

p. 75   Dams can become abodes . . . Rohr et al., 'Emerging', 449–50; Susanne H. Sokolow, Nicole Nova, Kim M. Pepin et al., 'Ecological Interventions to Prevent and Manage Zoonotic Pathogen Spillover', *Philosophical Transactions of the Royal Society B* 374 (2019), 6; Susanne Shultz, Hem Sagar Baral, Sheonaidh Charman et al., 'Diclofenac Poisoning Is Widespread in Declining Vulture Populations Across the Indian Subcontinent', *Proceedings of the Royal Society B* 271 (2004): S458–60; Wallace, *Big*, 128.

p. 75   As Rob Wallace and others have detailed . . . E.g. Rob Wallace, Alex Liebman, Luis Fernando Chaves and Rodrick Wallace, 'COVID-19 and Circuits of Capital', *Monthly Review*, 27 March 2020; Wallace, *Big*; Davis, *The Monster*; Fornace et al., 'Effects', 140–1; Jones, 'Zoonosis', 8401–3.

p. 76   With its copying of the US-style . . . Robert G. Wallace, Luke Bergmann, Lenny Hogerwerf and Marius Gilbert, 'Are Influenzas in Southern China Byproducts of the Region's Globalizing Historical Present?', in Gunn, *Influenza*, 101–44; Liu et al., 'Major', 68; Wu et al., 'Economic', 19–21.

p. 76   as the above-quoted German liberal pundit . . . Fücks, 'Weshalb'.

pp. 76–7   The pre-eminent philosopher of the plantation . . . John Locke, *Second Treatise of Government and a Letter Concerning Toleration* (Oxford:

Oxford University Press, 2016 [1689]), 26, 33. Emphasis in original. This argument draws on Andreas Malm, 'In Wildness Is the Liberation of the World: On Maroon Ecology and Partisan Nature', *Historical Materialism* 26 (2018): 3–37, and forthcoming work.

p. 79   read off from aggregate data . . . Erle C. Ellis, Kees Klein Goldewijk, Stefan Siebert et al., 'Anthropogenic Transformation of the Biomes, 1700 to 2000', *Global Ecology and Biogeography* 19 (2010): 589–606.

p. 79   One of the most remarkable calculated . . . Ali Alsamawi, Joy Murray and Manfred Lenzen, 'The Employment Footprint of Nations: Uncovering Master-Servant Relationships', *Journal of Industrial Ecology* 18 (2014): 59–70. Cf. e.g. Moana Simas, Richard Wood and Edgar Hertwich, 'Labor Embodied in Trade: The Role of Labor and Energy Productivity and Implications for Greenhouse Gas Emissions', *Journal of Industrial Ecology* 19 (2014): 343–55; and for a study explicitly measuring Marxian value transfers, Andrea Ricci, 'Unequal Exchange in the Age of Globalization', *Review of Radical Political Economics* 51 (2019): 225–45.

p. 80   Alf Hornborg has called it . . . The notion was developed in critical dialogue with David Harvey: see Alf Hornborg, *The Power of the Machine: Global Equalities of Economy, Technology, and Environment* (Walnut Creek: Altamira, 2001), 53. It was operationalised in Alf Hornborg, 'Footprints in the Cotton Fields: The Industrial Revolution as Time-Space Appropriation and Environmental Load Displacement', *Ecological Economics* 59 (2006): 74–81.

p. 80   This, as David Harvey has theorised it . . . David Harvey, *The Condition of Postmodernity: An Enquiry Into the Origins of Cultural Change* (Cambridge, MA: Blackwell, 1990), 240–1.

p. 81   In March 2020, climate . . . E.g. Friedman, 'With the Coronavirus'.

p. 81   'pretty much all potential . . . Steven Belmain quoted in Gruber, 'Predicting', 3–4.

p. 81   some of the very same forces . . . Johnson et al., 'Global', 2.

p. 82   One of the few scientific papers . . . Jamison Pike, Tiffany Bogich, Sarah Elwood et al., 'Economic Optimization of a Global Strategy to Address the Pandemic Threat', *PNAS* 111 (52), 18520.

p. 82   But as experts from Quammen to Wallace . . . The former quoted in Louise Boyle, ' "We Should Start Thinking About the Next One": Coronavirus Is Just the First of Many Pandemics to Come, Environmentalists Warn', *Independent*, independent.co.uk, 20 March 2020; Wallace et al., 'COVID-19'.

p. 82   One virologist announced . . . Brian Bird quoted in Vidal, 'Destroyed'. Another expert, Dennis Carroll of PREDICT, claimed that 'we're on a cycle of about every three years of getting something like this', i.e. something like SARS-CoV-2. Kevin Bergermarch, 'The Man Who Saw the Pandemic Coming', *Nautilus*, nautil.us, 12 March 2020.

p. 82   the eminent dialectical biologist . . . Levins et al., 'The Emergence', 55.

p. 82   'How much worse will . . . Dobson et al., 'Sacred Cows', 0718.

p. 82   'As the line dividing human . . . Gruber, 'Predicting', 1; Afelt et al., 'Bats, Coronaviruses', 2–3. Cf. e.g. Field, 'Bats', 282; Afelt et al., 'Bats, Bat-Borne', 118.

p. 82   The apathy in the sphere . . . E.g. Cunningham et al., 'One', 1, 3.

pp. 82–3   'unless there is a major global . . . Jones et al., 'Eco-Social', 32.

p. 83   'can we reverse or mitigate . . . Quammen, *Spillover*, 43.

p. 83   In an early analysis of Covid-19 . . . Wallace, 'Notes'. Emphasis in original. Cf. e.g. Wallace, *Big*, 280; Wallace and Wallace, *Neoliberal*, 82.

p. 83   the reaction of the Australian state apparatus to the bushfires . . . Livia Albeck-Ripka, Isabella Kwai, Thomas Fuller and Jamie Tarabay, ' "It's an Atomic Bomb": Australia Deploys Military as Fires Spread', *New York Times*, 5 January 2020 ('atomic bomb': Andrew Constance, transport minister in New South Wales).

pp. 84–5   simultaneous bad harvests . . . See e.g. Michelle Tigchelaar, David S. Battisti, Rosamond L. Naylor and Deepak K. Ray, 'Future Warming Increases Probability of Globally Synchronized Maize Production Shocks', *PNAS* 115 (2018): 6644–9; Timothy M. Lenton, Johan Rockström, Owen Gaffney et al., 'Climate Tipping Points – Too Risky to Bet Against', *Nature* 575 (2019): 592–5.

p. 85   the Pike paper estimates . . . Pike et al., 'Economic', 18521.

p. 86   In the first decade of the millennium, it accounted . . . Thomas F. Stocker, Dahe Qin, Gian-Kasper Plattner et al. (eds.) *Climate Change 2013: The Physical Science Basis. Contribution of Working Group I to the Fifth Assessment Report of the Intergovernmental Panel on Climate Change* (Cambridge: Cambridge University Press, 2013), 50; Florence Pendrill, U. Martin Persson, Javier Godar et al., 'Agricultural and Forestry Trade Drives Large Share of Tropical Deforestation Emissions', *Global Environmental Change* 56 (2019): 1–10; Henders et al., 'Trading'.

p. 86   Aviation remains a minor source . . . Rob Jordan, 'Global Carbon Emissions Growth Slows, but Hits Record High', Stanford Woods Institute for the Environment, news.stanford.edu, 3 December 2019; Gwyn Topham, 'Airlines' CO_2 Emissions Rising up to 70% Faster Than Predicted', *Guardian*, 19 September 2019.

p. 86   It has been well known for some time that global heating . . . For some seminal papers from early in the millennium, see Camille Parmesan and Gary Yohe, 'A Globally Coherent Fingerprint of Climate Change Impacts Across Natural Systems', *Nature* 421 (2003): 37–42; Terry L. Root, Jeff T. Price, Kimberly R. Hall et al., 'Fingerprints of Global Warming on Wild Animals and Plants', *Nature* 421 (2003): 57–60; Camille Parmesan, 'Ecological and Evolutionary Responses to Recent Climate Change', *Annual Review of Ecology, Evolution and Systematics* 37 (2006): 637–69.

p. 86   this process is unlikely to be gradual . . . Christopher H. Trisos, Cory Merow and Alex L. Pigot, 'The Projected Timing of Abrupt Ecological Disruption From Climate Change', *Nature* (2020), online first.

p. 87   Along the way, the animals meet . . . This paragraph is based on Colin J. Carlson, Gregory F. Albery, Cory Merow et al., 'Climate Change Will Drive Novel Cross-Species Viral Transmission', *BioRxiv*, biorxiv.org, 25 January 2020.

p. 88   Some act as vectors for infectious diseases . . . On this topic, see e.g. R. C. Andrew Thompson, 'Parasite Zoonoses and Wildlife: One Health, Spillover and Human Activity', *International Journal for Parasitology* 43 (2013), 1085; Paul E. Parham, Joanna Waldock, George K. Christophides et al., 'Climate, Environmental and Socio-Economic Change: Weighing Up the Balance in Vector-Borne Disease Transmission', *Philosophical Transactions of the Royal Society B* 370 (2015): 1–17; Diarmid Campbell-Lendrum, Lucien Manga, Magaran Bagayoko and Johannes Sommerfeld, 'Climate Change and Vector-Borne Diseases: What Are the Implications for Public Health Research and Policy?', *Philosophical Transactions of the Royal Society B* 370 (2015): 1–8; Joacim Rocklöv and Robert Dubrow, 'Climate Change: An Enduring Challenge for Vector-Borne Disease Prevention and Control', *Nature Immunology* (2020) online first.

p. 88   Similar uncertainties apply to locusts . . . Keith Cressman, 'Climate Change and Locusts in the WANA Region' in Mannava V. K. Sivakumar, Rattan Lal, Ramasamy Selvaraju and Ibrahim Hamdan (eds.), *Climate Change and Food Security in West Asia and North Africa* (Dordrecht: Springer, 2013), 131–43; Christine N. Meynard, Pierre-Emmanuel Gay, Michel Lecoq et al., 'Climate-Driven Geographic Distribution of the Desert Locust During Recession Periods: Subspecies' Niche Differentiation and Relative Risks Under Scenarios of Climate Change', *Global Change Biology* 23 (2017): 4739–49.

p. 89   'we cannot exclude a higher potential . . . Meynard et al., 'Climate-Driven', 4746.

p. 89   a second generation of swarms . . . Samuel Okiror, 'Second Wave of Locusts in East Africa Said to Be 20 Times Worse', *Guardian*, 13 April 2020 (Hellen Adoa quoted).

p. 89   Among mammals, bats have a special . . . Hayley A. Sherwin, W. Ian Montgomery and Mathieu G. Lundy, 'The Impact and Implications of Climate Change for Bats', *Mammal Review* 43 (2013): 171–82; Gareth Jones and Hugo Rebelo, 'Responses of Bats to Climate Change: Learning From the Past and Predicting the Future' in Adams and Pederson, *Bat*, 457–78.

p. 89   a raft of studies has reported . . . Richard K. LaVal, 'Impact of Global Warming and Locally Changing Climate on Tropical Cloud Forest Bats', *Journal of Mammalogy* 85 (2004): 237–44; Gary F. McCracken, Riley F. Bernard, Melquisidec Gamba-Rios et al., 'Rapid Range Expansion of the Brazilian Free-Tailed Bat in the Southeastern United States, 2008–2016', *Journal of Mammalogy* 99 (2018): 312–20; Ludmilla M. S. Aguiar, Enrico Bernard, Vivian Ribeiro et al., 'Should I Stay or Should I Go? Climate Change Effects on the Future of Neotropical Savannah Bats', *Global Ecology and Conservation* 5 (2016): 22–33; L. Ancillotto, L. Santini, N. Ranc et al., 'Extraordinary Range Expansion in a Common Bat: The Potential Roles of Climate Change and Urbanisation', *The Science of Nature* 103 (2016): 1–8.

p. 90   And China goes the same way . . . Jianguo Wu, 'Detection and Attribution of the Effects of Climate Change on Bat Distributions Over the Last 50 Years', *Climatic Change* 134 (2016): 681–96.

p. 90   then there are the modelling studies . . . Alice C. Hughes, Chutamas Satasook, Paul J. J. Bates et al., 'The Projected Effects of Climatic and Vegetation Changes on the Distribution and Diversity of Southeast Asian Bats', *Global Change Biology* 18 (2012): 1854–65. Cf. e.g. Hugo Rebelo, Pedro Tarroso and Gareth Jones, 'Predicted Impact of Climate Change on European Bats in Relation to Their Biogeographic Patterns', *Global Change Biology* 16 (2010): 561–76.

p. 90   It doesn't require any advanced mathematical . . . This implication is duly noted in e.g. Sherwin et al., 'The Impact', 11–12; Wu, 'Detection', 693. Most of the transmission and spillover events projected in Carlson et al., 'Climate', involve bats.

p. 90   Climate change is the supreme stressor . . . Jones and Rebelo, 'Responses', 460–1, 464–5; Sherwin et al., 'The Impact', 11; Plowright et al., 'Ecological', 5

p. 90   Some of the bats drifting to the Malaysian . . . Paul R. Epstein, Eric Chivian and Kathleen Frith, 'Emerging Diseases Threaten Conservation', *Environmental Health Perspectives* 111 (2003), 506.

p. 91   An exceptionally prolonged dry season . . . Bausch and Schwartz, 'Outbreak', 5.

p. 91   all three coronavirus epidemics so far . . . Sun et al., 'Potential', 3.

p. 91   In the last days of March 2020 . . . Janelle Griffith, 'Teen Whose Death May Be Linked to Coronavirus Denied Care for Not Having Health Insurance, Mayor Says', *NBC News*, nbcnews.com, 27 March 2020; Abigail Abrams, 'You Probably Read About an Uninsured Teen Who Died of COVID-19. The Truth Is More Complicated', *Time*, time.com, 2 April 2020; Holly Baxter, 'The Death of a 17-Year-Old and the Shame of a Healthcare System Which Wasn't Built to Work', *The Independent*, 7 April 2020.

p. 92   There were 12.5 intensive care . . . Figures from Adam Hanieh, 'This Is a Global Pandemic – Let's Treat It as Such', *Verso blog*, versobooks.com, 27 March 2020; Jackie Fox, 'African Countries Unprepared for the Storm', *RTE*, rte.ie, 2 April 2020.

p. 92   Prior health issues that weakened defences . . . Joseph A. McCartin, 'Class and the Challenge of COVID-19', *Dissent*, dissentmagazine.org, 23 March 2020; Jamie Smith Hopkins, 'A Likely but Hidden Coronavirus Risk Factor: Pollution', *Center for Public Integrity*, publicintegrity.org, 27 March 2020; Michael Sainato, ' "I've Already Got Infected Lungs": For Sick Coal Miners Covid-19 is a Death Sentence', *Guardian*, 19 April 2020.

p. 92   The rich could well afford . . . Jerod Davis, 'Coronavirus Diaries: I Charter Private Jets. Business Is Booming', *Slate*, slate.com, 8 March 2020; Rupert Neate, 'Super-Rich Jet Off to Disaster Bunkers Amid Coronavirus Outbreak', *Guardian*, 11 March 2020; Stacey Lastoe and Shivani Vora, 'The Wealthy Forge Ahead With (Slightly Altered) Travel Plans in Spite of "Stay at Home" Directives', *CNN*, cnn.com, 28 March 2020.

p. 92   Real estate agencies in the UK . . . Jonathan Prynn, 'The Rise of the "Corona Mansion": Wealthy Londoners Rushing to Let Sprawling Homes With Grounds for Up to £10k a Week Amid Covid-19 Lockdown', *Homes and Property*, homesandproperty.co.uk, 3 April 2020 (Ben Sloane of Aslon Chase quoted). The category of race is left out here; it is dealt with in a postscript to Malm and The Zetkin Collective, *White Skin*.

p. 93   Indeed, if the rich were first to be hit . . . Tom Phillips and Caio Barretto Briso, 'Brazil's Super-Rich and the Exclusive Club at the Heart of a Coronavirus Hotspot', *Guardian*, 4 April 2020.

p. 93   At the most formal, anaemic level . . . W. Neil Adger, 'Vulnerability', *Global Environmental Change* 16 (2006), 269; P. M. Kelly and W. N. Adger, 'Theory and Practice in Assessing Vulnerability to Climate Change and Facilitating Adaptation', *Climatic Change* 47 (2000), 328.

p. 94   In the post-war decades, research . . . See e.g. Kenneth Hewitt, 'The Idea of Calamity in a Technocratic Age', in Kenneth Hewitt (ed.), *Interpretations of Calamity: From the Viewpoint of Human Ecology* (London:

Allen and Unwin, 1983), 5–6; Hallie Eakin and Amy Lynd Luers, 'Assessing the Vulnerability of Social-Environmental Systems', *Annual Review of Environment and Resources* 31 (2006), 369; Christine Gibb, 'A Critical Analysis of Vulnerability', *International Journal of Disaster Risk Reduction* 28 (2018): 327–34.

p. 94 The first salvo was launched ... Ben Wisner, Phil O'Keefe and Ken Westgate, 'Global Systems and Local Disasters: The Untapped Power of People's Science', *Disasters* 1(1977): 47–57 (quotation from 48).

p. 95 This is the cardinal idea of critical ... Also known as 'structural' vulnerability theory; for an excellent overview, see Gibb, 'A Critical'.

p. 95 during a drought in northern Nigeria ... Michael Watts, 'On the Poverty of Theory: Natural Hazards Research in Context', in Hewitt, *Interpretations*, 258. See further this classical anthology; and for an application in the context of climate change, Hans G. Bohle, Thomas E. Downing and Michael J. Watts, 'Climate Change and Social Vulnerability: Toward a Sociology and Geography of Food Insecurity', *Global Environmental Change*, 4 (1994): 37–48.

p. 96 and this is exclusively 'determined ... Ben Wisner, Piers Blaikie, Terry Cannon and Ian Davis, *At Risk: Natural Hazards, People's Vulnerability and Disasters. Second Edition* (London: Routledge, 2005), 6, 91.

p. 96 One chapter in *At Risk* is devoted to this ... Ibid., 167–200.

p. 96 Meredeth Turshen attacked the paradigm ... Meredeth Turshen, 'The Political Ecology of Disease', *Review of Radical Political Economics* 9 (1977): 45–60. For most sophisticated essays in the same vein, see Richard Lewontin and Richard Levins, *Biology Under the Influence: Dialectical Essays on Ecology, Agriculture, and Health* (New York: Monthly Review Press, 2007).

p. 96 'The doctors can trace the fatal infection ... J. P. Nettl, *Rosa Luxemburg: The Biography* (London: Verso, 2019 [1966]), 478.

p. 97 Since the 1970s, critical epidemiology has agreed ... For applications to zoonoses, see e.g. Paul Farmer, 'Social Inequalities and Emerging Infectious Diseases', *Emerging Infectious Diseases* 2 (1996): 259–69; Vupenyu Dzingirai, Salome Bukachi, Melissa Leach et al., 'Structural Drivers of Vulnerability to Zoonotic Disease in Africa', *Philosophical Transactions of the Royal Society B* 372 (2017): 1–9.

p. 97 Wisner et al. have made it enormously influential ... Wisner et al., *At Risk*, 49–84.

p. 97 'revolution or major realignment ... Ibid., 91. Cf. Ben Wisner, 'Flood Prevention and Mitigation in the People's Republic of Mozambique', *Disasters* 3 (1979), 305.

p. 97 **'must be, broadly speaking ...** Paul Susman, Phil O'Keefe and Ben Wisner, 'Global Disasters, a Radical Interpretation' in Hewitt, *Interpretations*, 280.

p. 99 **'much mystifying argument about ...** Susman et al., 'Global', 265. The scepticism was already there in Wisner et al., 'Global', 48.

pp. 99–100 **'Most disasters, or more correctly ...** Geoff O'Brien, Phil O'Keefe, Joanne Rose and Ben Wisner, 'Climate Change and Disaster Management', *Disasters* 30 (2006), 65. Emphasis added.

p. 100 **'With climate change, human action ...** Wisner et al. 2005, *At Risk*, 121. See further e.g. 83, 114, 149, 195–6, 213. Emphasis in original.

p. 100 **Extending the artwork from Wisner et al., we might picture it ...** The left side of this model sums up some of the findings in Andreas Malm and Shora Esmailian, 'Ways In and Out of Vulnerability to Climate Change: Abandoning the Mubarak Project in the Northern Nile Delta, Egypt', *Antipode* 45 (2013): 474–92; Andreas Malm and Shora Esmailian, 'Doubly Dispossessed by Accumulation: Egyptian Fishing Communities Between Enclosed Lakes and a Rising Sea', *Review of African Political Economy* (2012): 408–26; Andreas Malm, 'Sea Wall Politics: Uneven and Combined Protection of the Nile Delta Coastline in the Face of Sea Level Rise', *Critical Sociology* 39 (2013): 803–32. The right side builds on Andreas Malm, *Fossil Capital: The Rise of Steam Power and the Roots of Global Warming* (London: Verso, 2016).

p. 103 **'It is clear that people in Italy ...** Ingar Solty, 'The Bio-Economic Pandemic and the Western Working Classes', *The Bullet*, socialistproject.ca, 24 March 2020.

p. 104 **'Let's stop the outbreaks ...** Wallace, 'Notes'. Cf.: 'Neoliberal and other kinds of exploitation can scramble long-selected patterns and processes associated with said protection [against zoonotic diseases] even when "public health" measures are kept in place. That interference alone may suffice to trigger explosive outbreaks of even previously marginalized vector-borne infections or ratchets in public health degradation.' Wallace et al., *Clear-Cutting*, 56. Quammen made the same general point – without any analysis of neoliberalism or other root causes – in David Quammen, 'We Made the Coronavirus Epidemic', *New York Times*, 28 January 2020. He destroys the view of zoonotic spillovers as acts of God in *Spillover*, e.g. 36.

p. 104 **'suggest the need for a complete ...** Nick Paton Walsh and Vasco Cotovio, 'Bats Are Not to Blame for Coronavirus. Humans Are', *CNN*, 20 March 2020.

p. 104 **Some people under lockdown ...** Michael Marshall, 'Covid-19 – a Blessing for Pangolins?', *Guardian*, 18 April 2020.

p. 106 **Deep in the Amazon, the Brazilian oil** ... Lucia Greyl and Camila Rolando Mazzuca, 'Gas Pipeline Urucu-Coari-Manaus and Urucu–Porto Velho – Petrobras, Brazil', *Environmental Justice Atlas*, ejatlas.org, 21 January 2016 (for more examples of fossil fuel extraction in tropical forests, see this eminent site); Petrobras, 'Solimões Basin', petrobras.com.br, n. d; *Brazil Energy Insight*, 'Rosneft Plans New Wells and Pipelines in the Solimões Basin', brazilenergyinsight.com, 21 June 2019; Kevin Koenig and Amazon Watch, *The Amazon Sacred Headwaters: Indigenous Rainforest 'Territories for Life' Under Threat*, amazonwatch.org, 9 December 2019.

p. 106 **On the other side of the tropics, in Indonesian Sumatra** ... Elviza Diana, 'A Forest Beset by Oil Palms, Logging, Now Contends With a Coal-Trucking Road', *Mongabay*, news.mongabay.com, 28 May 2019; Elviza Diana, 'In Sumatra, an Indigenous Plea to Stop a Coal Road Carving Up a Forest', *Mongabay*, 8 April 2020.

p. 107 **This swampy rainforest area** ... Quammen, *Spillover*, e.g. 37; Greta C. Dargie, Simon L. Lewis, Ian T. Lawson et al., 'Age, Extent and Carbon Storage of the Central Congo Basin Peatland Complex', *Nature* 542 (2017): 86–90; Phoebe Weston, 'Plan to Drain Congo Peat Bog for Oil Could Release Vast Amount of Carbon', *Guardian*, 28 February 2020.

3. War Communism

p. 109 **the Bank of England has just flagged a 'contraction** ... Jasper Jolly and Graeme Wearden, 'Bank of England Warns of Worst Contraction in Centuries, as Economic Activity Slumps – Business Live', *Guardian*, 23 April 2020. Emphasis added.

p. 110 **When capital runs in the spiral** ... James O'Connor, 'Capitalism, Nature, Socialism: A Theoretical Introduction', *Capitalism Nature Socialism* 1 (1988), 11–38; 'On the Two Contradictions of Capitalism', *Capitalism Nature Socialism* 2 (1991), 107–9; both included alongside further reflections in *Natural Causes: Essays in Ecological Marxism* (New York: Guilford Press, 1998).

p. 110 **But then, O'Connor continues** ... O'Connor, 'Capitalism', 11. Emphasis in original. O'Connor is here heavily indebted to Karl Polanyi: see Alan P. Rudy, 'On Misunderstanding the Second Contradiction Thesis', *Capitalism Nature Socialism* 30 (2019): 17–35.

p. 111 **Nancy Fraser, who has developed similar ideas** ... Nancy Fraser and Rahel Jaeggi, *Capitalism: A Conversation in Critical Theory* (Cambridge: Polity, 2018), e.g. 36, 94.

p. 111 'self-destruct by impairing or destroying . . . O'Connor, 'Capitalism', 25.

p. 111 'it is not true that the current era . . . Martin Spence, 'Capital Against Nature: James O'Connor's Theory of the Second Contradiction of Capitalism', *Capital and Class* 24 (2000), 93–4.

p. 111 'internal grammar' of nature . . . Fraser and Jaeggi, *Capitalism*, 37. Wallace predicts a crisis from zoonotic spillover as a manifestation of the second contradiction in *Big*, 59.

p. 112 global warming would suppress profits . . . See e.g. O'Connor, 'Capitalism', 25.

p. 112 O'Connor himself hinted . . . E.g. ibid., 23.

p. 112 expected that they would have opposite signatures . . . O'Connor, 'On the Two', 108.

p. 113 'The experience of our generation . . . Walter Benjamin, *The Arcades Project* (Cambridge, MA: Harvard University Press, 2002), 667.

p. 114 In *The Fate of Rome* . . . Kyle Harper, *The Fate of Rome: Climate, Disease, and the End of an Empire* (New Haven: Yale University Press, 2017) (quotations from 2–3).

p. 115 'armed climate refugees . . . Ibid., 192.

pp. 115–16 'the combination of plague and climate . . . Ibid., 245.

p. 116 'A precociously global world . . . Ibid., 293.

p. 116 a rebuttal of *The Fate* in three instalments . . . John Haldon, Hugh Elton, Sabine R. Heubner et al., 'Plagues, Climate Change, and the End of an Empire: A Response to Kyle Harper's *The Fate of Rome* (1): Climate', *History Compass* 16 (2018): 1–13; '(2): Plagues and a Crisis of Empire', 1–10; '(3): Disease, Agency, and Collapse', 1–10.

p. 117 'Simply assuming a causal connection . . . Haldon et al., '(1): Climate', 5.

p. 117 This has recently become a temptation . . . This draws on the very judicious survey and critique of the field in Guy D. Middleton, *Understanding Collapse: Ancient History and Modern Myths* (Cambridge: Cambridge University Press, 2017).

p. 118 On the contrary, lives of working people in Egypt . . . This draws on forthcoming work.

p. 118 The popular relief from ancient collapse . . . See Middleton, *Understanding*.

p. 120 Rosa Luxemburg very famously objected . . . For an account of the revisionism debate, see the magnificent Nettl, *Rosa*, especially 202–50.

p. 122 'responsible driving and civic . . . James C. Scott, *Two Cheers for Anarchism* (Princeton: Princeton University Press, 2012), 82.

p. 123 The early weeks of the pandemic ... See e.g. Covid-19 Mutual Aid UK, covidmutual.org; Tom Anderson, 'An Inside Look at One of Bristol's New Coronavirus Mutual Aid Groups', *The Canary*, thecanary.co, 28 March 2020; *It's Going Down*, 'Autonomous Groups Are Mobilizing Mutual Aid Initiatives to Combat the Coronavirus', itsgoingdown.org, 20 March 2020; Woodbine, 'From Mutual Aid to Dual Power in the State of Emergency', *Roar*, roarmag.org, 22 March 2020; Eleanor Goldfield, 'Mutual Aid: "When the System Fails, the People Show Up" ', *Roar*, 22 April 2020.

p. 123 Some caught a glimpse of ... Notably George Monbiot, 'The Horror Films Got It Wrong. This Virus Has Turned Us All Into Caring Neighbours', *Guardian*, 31 March 2020.

p. 123 In the favelas of Rio de Janeiro ... Caio Barretto Briso and Tom Phillips, 'Brazil Gangs Impose Strict Curfews to Slow Coronavirus Spread', *Guardian*, 25 March 2020.

p. 124 Revolutionaries remained outside the structures ... Asef Bayat, *Revolution without Revolutionaries: Making Sense of the Arab Spring* (Stanford: Stanford University Press, 2017), 169. Emphasis added.

p. 125 In the second week of September 1917, Lenin ... V. I. Lenin, *Revolution at the Gates: Selected Writings From February to October 1917* (London: Verso, 2002 [1917]), 69, 103.

p. 125 As if that were not enough, heavy floods ... Lars T. Lih, *Bread and Authority in Russia, 1914–1921* (Berkeley: University of California Press, 1990), 65–7.

pp. 125–6 'Famine, genuine famine' ... N. Dolinsky quoted in ibid., 111.

p. 126 'We are nearing ruin ... Lenin, *Revolution*, 70, 91, 129.

p. 126 Paris to Petrograd had 'outlined, determined ... Ibid., 74, 111, 46. Emphases in original.

p. 127 Against the Kerensky government's feeble ... Ibid., 40, 74. Emphases in original.

p. 129 There is an urgent need to stop deforestation ... Ghulam Nabi, Rabeea Siddique, Ashaq Ali and Suliman Khan, 'Preventing Bat-Borne Viral Outbreaks in Future Using Ecological Interventions', *Environmental Research* (2020), online first.

p. 129 'The most effective way to prevent ... Jie Cui et al., 'Origin', 190.

p. 129 'The most effective place to address ... Wood et al., 'A Framework', 2882. Cf. e.g. Keesing et al., 'Impacts', 651; Cunningham et al., 'One Health', 5–6.

p. 129 'Allocation of global resources ... Pike et al., 'Economic', 18522.

p. 131 Given what we know about bats ... Cf. Hughes et al., 'The Projected', 1863; Sokolow et al., 'Ecological', 2, 6; Schneeberger and Voigt, 'Zoonotic', 279, 282.

p. 130 **The immediate beneficiaries will be** . . . Cf. Walsh et al., 'Forest', 1812.

p. 130 **It would give some room back** . . . See e.g. Rohr et al., 'Emerging', 451–3.

p. 131 **'We need (for a certain transitional period** . . . Lenin, *Revolution*, 41; Wallace and Wallace, *Neoliberal*, 90. Emphasis in original.

p. 131 **Between 2004 and 2012, deforestation in** . . . Doug Boucher, Sarah Roquemore and Estrellita Fitzhugh, 'Brazil's Success in Reducing Deforestation', *Tropical Conservation Science* 6 (2013): 426–45; Joana Castro Pereira and Eduardo Viola, 'Catastrophic Climate Risk and Brazilian Amazonian Politics and Policies: A New Research Agenda', *Global Environmental Politics* 19 (2019): 93–103; Thomas M. Lewinsohn and Paulo Inácio Prado, 'How Many Species Are There in Brazil?', *Conservation Biology* 19 (2005): 619–24. On how radically Brazil departed from the trends in these years, see e.g. Austin et al., 'Trends'; Lambin and Meyfroidt, 'Global', 3467.

p. 132 **'Rosa Luxemburg has a great line about revolution** . . . Rodrigo Nunes in Juan Grigera, Jeffery R. Webber, Ludmila Abilio et al., 'The Long Brazilian Crisis: A Forum', *Historical Materialism* 27 (2019), 72.

p. 132 **After SARS, the state took some perfunctory** . . . Nian Yang, Peng Liu, Wenwen Li and Li Zhang, 'Permanently Ban Wildlife Consumption', *Science* 367 (2020): 1434–5 (quotation from 1435). Cf. Felbab-Brown, *The Extinction*, e.g. 15.

p. 133 **Jingjing Yuan and colleagues went a step further** . . . Yuan et al., 'Regulating', 2–3.

p. 133 **research indicates that awareness** . . . Li et al., 'Human-Animal', 87.

p. 133 **online sellers touted medicines** . . . Rahel Nuwer, 'To Prevent Next Coronavirus, Stop the Wildlife Trade, Conservationists Say', *New York Times*, 19 February 2020.

p. 133 **It has been argued that the moral** . . . E.g. Brian Barth, 'The Surprising History of the Wildlife Trade That May Have Sparked the Coronavirus', *Mother Jones*, motherjones.com, 29 March 2020.

p. 134 **One of the few success stories** . . . Felbab-Brown, *The Extinction*, 234–8 (quotation from 237). On the normative effects of prohibition, cf. 105.

p. 135 **Some have argued that a blanket abolition** . . . See e.g. Smriti Mallapaty, 'China Set to Clamp Down Permanently on Wildlife Trade in Wake of Coronavirus', *Nature*, nature.com, 21 February 2020; Robert G. Webster, 'Wet Markets – A Continuing Source of Severe Acute Respiratory Syndrome and Influenza?', *The Lancet* 363 (2004), 236; Jane Qiu, 'How China's'.

p. 135 **Even Germany has been identified** . . . Sarah Heinrich, Arnulf Koehncke and Chris R. Shepherd, 'The Role of Germany in the Illegal Global Pangolin Trade', *Global Ecology and Conservation* 20 (2019): 1–7.

p. 135 **Barack Obama purported to make . . .** Felbab-Brown, *The Extinction*, 110.

p. 135 **But law enforcement would . . .** Ibid., 27, 107–8, 112, 115–16, 245, 260.

p. 136 **The main alternative to such an approach is to legalise . . .** Ibid., 50, 105, 145–6, 243, 247–8; quotation from 116.

p. 136 **One would wish that lifting . . .** E.g. Rogen, 'Socioeconomic', 25, 29; van Vliet and Mbazza, 'Recognizing', 49–51; Felbab-Brown, *The Extinction*, 52.

pp. 136–7 **It has, on the other hand, been vociferously argued . . .** Cf. Nathalie van Vliet, ' "Bushmeat Crisis" and "Cultural Imperialism" in Wildlife Management? Taking Value Orientations Into Account for a More Sustainable and Culturally Acceptable Wildmeat Sector', *Frontiers in Ecology and Evolution* 6 (2018): 1–6; Felbab-Brown, *The Extinction*, 26, 51.

p. 137 **Unfortunately, that argument is self-defeating . . .** Cf. Ripple et al., 'Bushmeat', 8.

p. 137 **'incentives for communities to switch . . .** Ibid., 10–11. Cf. e.g. Poulsen et al., 'Bushmeat'; Pike et al., 'Economic', 18522.

p. 137 **And, so, one could look to Cuba, which seems . . .** Patrick Oppmann, 'Coronavirus-Hit Countries Are Asking Cuba for Medical Help. Why Is the US Opposed?', *CNN*, 26 March 2020; Helen Yaffe, 'The World Rediscovers Cuban Medical Internationalism', *Le Monde Diplomatique*, mondediplo.com, 30 March 2020; Peter Kornbluh, 'Cuba's Welcome to a Covid-19-Stricken Cruise Ship Reflects a Long Pattern of Global Humanitarian Commitment', *The Nation*, thenation.com, 21 March 2020; and the superb analysis in Robert Hush, 'Why Does Cuba "Care" So Much? Understanding the Epistemology of Solidarity in Global Health Outreach', *Public Health Ethics* 7 (2014): 261–76.

p. 138 **'If anything real is to be done . . .** Lenin, *Revolution*, 81; cf. 82, 101, 128.

p. 139 **This begins with a nationalisation of all . . .** The idea of nationalising fossil fuel companies bubbled up on the left fringe of the debates in Covid-19: e.g. Damian Carrington, Jillian Ambrose and Matthew Taylor, 'Will the Corona Crisis Kill the Oil Industry and Help Save the Climate?', *Guardian*, 1 April 2020; Chris Saltmarsh, 'Do Not Resuscitate the Oil Industry', *Tribune*, tribunemag.co.uk, 21 April 2020. Cf. e.g. Peter Gowan, 'A Plan to Nationalize Fossil-Fuel Companies', *Jacobin*, jacobinmag.com, 26 March 2018.

p. 140 **With machines that look like large fans . . .** The arguments in these paragraphs draw on forthcoming work together with Wim Carton, a pioneer of this line of research. Sources used here: field observations at Climeworks in Zürich by Wim Carton and the author in February 2020; Christoph Beuttler, Louise Charles and Jan Wurzbacher, 'The Role of Direct Air Capture in Mitigation of Anthropogenic Greenhouse Gas Emissions', *Frontiers in Climate* 1 (2019): 1–7; Giulia Realmonte, Laurent Drouet, Ajay Gambhir et al., 'An Inter-Model Assessment of the Role of Direct Air Capture in Deep Mitigation Pathways', *Nature Communications* 10 (2019): 1–12; Mahdi Fasihi, Olga Efimova and Christian Breyer, 'Techno-Economic Assessment of CO_2 Direct Air Capture Plants', *Journal of Cleaner Production* 224 (2019): 957–80; Yuki Ishimoto, Masahiro Sugiyama, Etsushi Kato et al., 'Putting Costs of Direct Air Capture in Context', *Forum for Climate Engineering Assessment Working Paper Series*, School of International Service, American University, June 2017.

p. 140 BECCs would devour such monstrous … See e.g. Michael Obersteiner, Johannes Bednar, Fabian Wagner et al., 'How to Spend a Dwindling Greenhouse Gas Budget', *Nature Climate Change* 8 (2018), 8; Christopher B. Field and Katharine J. Mach, 'Rightsizing Carbon Dioxide Removal', *Science* 356 (2017), 707; Kate Dooley, Peter Christoff and Kimberley A. Nicholas, 'Co-Producing Climate Policy and Negative Emissions: Trade-Offs for Sustainable Land-Use', *Global Sustainability* 1 (2018), 1; Vera Heck, Dieter Gerten, Wolfgang Lucht and Alexander Popp, 'Biomass-Based Negative Emissions Difficult to Reconcile With Planetary Boundaries', *Nature Climate Change* 8 (2018): 151–5.

p. 140 because they are small and easily switched … Jan Wohland, Dirk Witthaut and Carl-Friedrich Schleussner, 'Negative Emission Potential of Direct Air Capture Powered by Renewable Excess Electricity in Europe', *Earth's Future* 6 (2018): 1380–4.

p. 141 it could go into microalgae or liquid fuel … Jennifer Wilcox, Peter C. Psarras and Simona Liguori, 'Assessment of Reasonable Opportunities for Direct Air Capture', *Environmental Research Letters* 12 (2017): 1–7.

p. 141 'That CO_2 could then be pressurized … Jeff Tollefson, 'Price of Sucking CO_2 From Air Plunges', *Nature* 558 (2018), 173.

pp. 141–2 As Holly Jean Buck shows in *After Geoengineering* … Holly Jean Buck, *After Geoengineering: Climate Tragedy, Repair, and Restoration* (London: Verso, 2019), e.g. 32–3, 122, 126–9, 191. Readers interested in the topic could then peruse J. P. Sapinski, Holly Jean Buck and Andreas Malm (eds.), *Has It Come to This? The Promises and Perils of Geoengineering on the Brink* (New Brunswick: Rutgers University Press, 2020).

p. 142 Other students of direct air capture … See e.g. Realmonte et al., 'An Inter-Model', 6.

p. 143 'getting rid of these corporations . . . Buck, *After*, 136; cf. 186, 203, 206.

p. 144 In late April 2020, *Scientific American* publicised . . . Benjamin Storrow, 'Why CO₂ Isn't Falling More During a Global Lockdown', *Scientific American*, 24 April 2020.

p. 144 'a single economic plan covering the whole country . . . Leon Trotsky, *Terrorism and Communism* (London: Verso, 2007 [1920]), 147.

p. 145 'The ways of combating catastrophe . . . Lenin, *Revolution*, 70, 74. Emphasis in original.

p. 146 in mid-April 2020, one of the largest experiments . . . Graham Readfearn, 'Scientists Trial Cloud Brightening Equipment to Shade and Cool Great Barrier Reef', *Guardian*, 16 April 2020.

p. 147 As one of the sharpest scholars . . . Kevin Surprise, 'Preempting the Second Contradiction: Solar Geoengineering as Spatiotemporal Fix', *Annals of the American Association of Geographers* 108 (2018): 1228–44. There is a clear risk that capital deploys negative emissions technologies, including direct air capture, in an analogous manner: see Wim Carton, ' "Fixing" Climate Change by Mortgaging the Future: Negative Emissions, Spatiotemporal Fixes, and the Political Economy of Delay', *Antipode* 51 (2019): 750–69.

p. 148 It is worth re-emphasising . . . This and the following paragraphs draw on Norman Geras, *The Legacy of Rosa Luxemburg* (London: Verso, 1983 [1976]), 14–39. Luxemburg quotations are from 21, 32–4.

p. 150 the fight ends 'either in . . . Karl Marx and Friedrich Engels, *The Communist Manifesto: A Modern Edition* (London: Verso, 2012 [1848]), 35.

p. 150 'conditional mood of the probability . . . Daniel Bensaïd, *An Impatient Life* (London: Verso, 2013), 291.

p. 150 'Whether the probable disaster can be avoided . . . Daniel Bensaïd, ' "Leaps! Leaps! Leaps!" ' in Sebastian Budgen, Stathis Kouvelakis and Slavoj Žižek (eds.), *Lenin Reloaded: Toward a Politics of Truth* (Durham: Duke University Press, 2007), 159; Lenin, *Revolution*, 155.

p. 151 Nothing can now be saved . . . Lenin, *Revolution*, 157.

p. 152 'The deeper the crisis, the more . . . Georg Lukács, *Lenin: A Study in the Unity of His Thought* (London: Verso, 2009 [1924]), 29.

p. 152 The 2013 edition of the 'worldwide threat . . . 'US Intelligence Community Worldwide Threat Assessment, Statement for the Record, March 12, 2013' in *United States Central Intelligence Agency (CIA) Handbook: Strategic Information, Activities and Regulations* (Washington, DC: International Business Publications, 2013), 40.

p. 152 'sums up Leninist politics . . . Bensaïd, ' "Leaps!" ', 153.

pp. 152–3 'set to work to stir up all and sundry . . . Quoted in ibid., 156–7.

p. 153 'The inherent tendencies of capitalist development . . . Quoted in Geras, *The Legacy*, 34–5.

p. 154 Peace is a better thing than war . . . Johannes Kester and Benjamin K. Sovacool, 'Torn Between War and Peace: Critiquing the Use of War to Mobilize Peaceful Climate Action', *Energy Policy* 104 (2017): 50–5. Cf. Laurence L. Delina and Mark Diesendorf, 'A Response to Kester and Sovacool', *Energy Policy* 112 (2018): 1–3. For a similar argument against martial metaphors in climate politics, see Kate Yoder, 'War of Words', *Grist*, 5 December 2018.

p. 155 The question is rather, as Alexandria Ocasio-Cortez . . . Her Twitter account, 6 January 2020.

p. 155 not as tangible as aerial bombardment . . . Kester and Sovacool, 'Torn', 52.

p. 156 psychological research showing that ordinary Americans . . . Stephen J. Flusberg, Teenie Matlock and Paul H. Thibodeau, 'Metaphors for the War (or Race) Against Climate Change', *Environmental Communication* 11 (2017): 769–83.

p. 157 the Spanish flu, introduced by . . . Afkhami, 'Compromised', 373; Sumiko Otsubo, 'Fighting on Two Fronts: Japan's Involvement in the Siberian Intervention and the Spanish Influenza Pandemic of 1918' in Tosh Minohara, Tze-ki Hon and Ewan Dawley (eds.), *The Decade of the Great War: Japan and the Wider World in the 1910s* (Leiden: Brill, 2014), 461–80.

p. 157 'Either the lice will defeat socialism . . . Quoted in S. A. Smith, *Russia in Revolution: An Empire in Crisis, 1890–1928* (Oxford: Oxford University Press, 2017), 162; cf. 218, 232, 245.

p. 157 'Through the interminable nights . . . Victor Serge, *Year One of the Russian Revolution* (London: Bookmarks and Pluto, 1992 [1930]), 366.

p. 158 In the half-year before the Civil War . . . Silvana Malle, *The Economic Organization of War Communism 1918–1921* (Cambridge: Cambridge University Press, 1985), 49–68; Smith, *Russia*, 220–2, 233, 249.

p. 158 'the incredibly rapid way in which the privileged . . . Smith, *Russia*, 236.

p. 159 Silvana Malle notes that the territory . . . Malle, *The Economic*, 396, 63, 219–20.

p. 159 'An industry which is completely deprived . . . Trotsky, *Terrorism*, 122.

p. 160 This poor substitute had provided . . . Malle, *The Economic*, 64.

p. 160 'we had to stoke our boilers . . . Trotsky, *Terrorism*, 122.

p. 160 Malle describes how the state set up . . . Malle, *The Economic*, 220–6, 502.

p. 161   'Our fuel requirements cannot be satisfied . . . Trotsky, *Terrorism*, 126, 129, 10; see further 145–7; Malle, *The Economic*, 479–80, 485–6, 502.

p. 161   'was not extensively implemented . . . Malle, *The Economic*, 502.

p. 162   'An habitual, normal regime . . . Quoted in Lars T. Lih, ' "Our Position Is in the Highest Degree Tragic": Bolshevik "Euphoria" in 1920' in Mike Haynes and Jim Wolfreys (eds.), *History and Revolution: Refuting Revisionism* (London: Verso, 2007), 121.

p. 162   'are no more images of utopia than . . . Terry Eagleton, 'Lenin in the Postmodern Age' in Budgen et al., *Lenin*, 48.

p. 164   'The earth the wretched would . . . Salvage Editorial Collective, 'Tragedy of the Worker: Toward the Proletarocene', *Salvage* no. 7 (2019), 60, 55.

p. 164   'our position is in the highest degree . . . Lih, ' "Our Position" '.

p. 164   as Lars T. Lih has showed in a series . . . Lars T. Lih, 'Deferred Dreams: War Communism 1918–1921', *The National Council for Soviet and East European Research*, Washington, D.C., 8 May 1995; Lars Lih, 'Bukharin's "Illusion": War Communism and the Meaning of NEP', *Russian History* 27 (2000): 417–59; Lih, ' "Our Position" '.

p. 165   'is bound to draw from the old institutions . . . Trotsky, *Terrorism*, 113.

p. 165   'the cure may be worse . . . Kester and Sovacool, 'Torn', 52.

p. 166   'oddly libertarian Leninism' . . . Bensaïd, *An Impatient*, 317.

p. 167   'an analogy very rich in content . . . Trotsky, *Terrorism*, 133.

p. 168   Two days after seizing power, the Bolsheviks . . . For this history, see Douglas R. Weiner, *Monuments of Nature: Ecology, Conservation, and Cultural Revolution in Soviet Russia* (Pittsburgh: University of Pittsburgh Press, 2000).

p. 168   At the height of the war in 1919, Bolshevik activists . . . This story is told in J. Veselý, 'Vladimir Iljic Lenin and the Conservation of Nature', *Zoologické Listy* 18–19 (1970), 19–20, 197–8.

p. 168   bringing to the fore such figures as . . . Weiner, *Monuments*, 35, 37.

p. 169   Russians were first to propose . . . Ibid., viii. Emphasis added.

p. 170   *New York Times* ran an opinion piece expressing some admiration . . . Fred Strebeigh, 'Lenin's Eco-Warriors', *New York Times*, 7 August 2017. Note that we are here leaving out the thorny question of what methods can best conserve wild nature. A systematic investigation of approaches to wilderness is part of a forthcoming work.

p. 170   'There is a universal feeling . . . Theodor Adorno, *Negative Dialectics* (New York: Bloomsbury, 2007 [1966]), 67.

p. 171   Nature is today more than ever conceived . . . Max Horkheimer, *Eclipse of Reason* (New York: Oxford University Press, 1947), 108–9. For one readable recent application of the 'domination of nature' theory to the fate of domesticated and wild animals, see Diana Stuart and Ryan Gunderson, 'Human-Animal Relations in the Capitalocene: Environmental Impacts and Alternatives', *Environmental Sociology* 6 (2020): 68–81.

p. 172   **Particularly nervous about primates . . .** Thomas R. Gillespie and Fabian H. Leendertz, 'Great-Ape Health in Human Pandemics', *Nature* 579 (2020): 497.

p. 172   **Infectious diseases have already annihilated . . .** Magdalena Bermejo, José Domingo Rodríguez-Teijeiro, Germán Illera et al., 'Ebola Outbreak Killed 5000 Gorillas', *Science* 314 (2006), 1564; Cunningham et al., 'One', 2.

p. 173   **This episode in the ecological crisis . . .** Cf. Quammen, *Spillover*, 388; Wallace, *Big*, 34; and for the general argument, Andreas Malm, *The Progress of This Storm: Nature and Society in a Warming World* (London: Verso, 2018).

p. 173   **'human beings become conscious of . . .** Theodor Adorno, *History and Freedom: Lectures 1964–1965* (Cambridge: Polity, 2006), 151-2.

p. 174   **whether humanity is capable of preventing . . .** Theodor W. Adorno, *Critical Models: Interventions and Catchwords* (New York: Columbia University Press, 2005), 144.